Security, Race, Biopower

Holly Randell-Moon • Ryan Tippet
Editors

Security, Race, Biopower

Essays on Technology and Corporeality

Editors
Holly Randell-Moon
Department of Media, Film and
Communication
University of Otago
Dunedin, New Zealand

Ryan Tippet
Department of Media, Film and
Communication
University of Otago
Dunedin, New Zealand

ISBN 978-1-349-71670-8   ISBN 978-1-137-55408-6   (eBook)
DOI 10.1057/978-1-137-55408-6

Library of Congress Control Number: 2016956421

Cover illustration: © YAY Media AS / Alamy Stock Photo

Printed on acid-free paper

This Palgrave Macmillan imprint is published by Springer Nature
The registered company is Macmillan Publishers Ltd. London

Introduction

Security, Race, Biopower

This is a book about technologies. *Security, Race, Biopower* explores the global abundance of technologies in medicine, media, surveillance, and war that are used to target and extend the lives of people differently in different geographical locations. The contributors show how technologies of population management—the ways in which bodies and lives are moulded to the benefit of governing authorities—are connected to historical and contemporary forms of racism that justify geographical and social inequalities. The book contends that the application and dissemination of contemporary technologies is premised on an economisation of these resources in favour of those who "deserve" life, based on the space and race to which a body belongs. We argue that the theories of French philosopher Michel Foucault on biopower and security are critical to understanding these technological determinations of deserving life. Biopower is broadly defined by Foucault as 'a technology of power centered on life' (1991, p. 266), which is characteristic of the modern Western state. The term encapsulates a central concern of governing authorities: how can we foster the health and wellbeing of citizens so they can live and therefore work longer? Enacting strategies to address this question does not presume all bodies within a population are of equal "value". Because the state has a limited amount of resources to produce good health, bio-power works in a distributive manner: 'to qualify, measure, appraise, and hierarchize' (p. 266). Racism and race are bound up in these hierarchies of valuable life and the distributive application of technologies to bodies and space. Contributions to this book show, for instance, how racism explains why some bodies become the target for drone kills, while others are targeted as ideal consumers of drones as toys; why some bodies receive human immunodeficiency virus (HIV) preventative medicine and others do not; why some bodies can consume iPhones while the bodies of those who make them suffer toxic disease and distress. Such inequalities in the technological lives of the global population do not go uncontested—indeed they are the subject of significant political and social protest. This book argues that processes of racialisation, attaching race to bodies in particular spatial locations, services an economic and geopolitical situation wherein citizens extend and improve their own lives at the expense of others, whose bodies are deemed disposable or surplus to population needs.

Biopower's concern with securing the health and wellbeing of the population through careful management and calculation of those bodies within it serves as an overriding theme for the various contributions and case studies offered in the book. Foucault conceives of biopower operating on two levels—manipulating bodies *as individuals* with discipline and *as species* within biopolitics. There is a particular focus in the final section of the book, where the previous examinations of geography, technology, racisms, and life are bought together, on biopolitics: the state's comprehension and grasp of the population as a 'biological problem' (Foucault 2003, p. 245). Biopolitical strategies address a 'global mass that is affected by overall processes characteristic of birth, death, production, illness, and so on' (pp. 242–3). Biopolitics necessarily institutes divisions—or 'caesurae'—within the population, drawing upon an evolutionist logic which valorises certain bodies and delegitimises others in order to protect the "purity" and health of the population overall. Along with biopower and its attendant topic biopolitics, this book offers two other critical concepts to explain the intersections of racism, geography, and technologies of security and health: geocorpographies and somatechnics.

Developed by Joseph Pugliese to explain how the treatment of Iraqi and Afghan civilians in the "War on Terror" is influenced by the geographical locations they inhabit (2007), geocorpographies combines geography with corporeality, space with bodies. Here geography is not a neutral backdrop to the ways bodies can move, occupy, or use the space they are in. Rather, bodies and bodily identity are always already constrained or enabled by their placement in space. Think of how one's identity as a mother, a worker, or an object of sexual attention can change in different spaces and change the movement of the body through that space. More critically, and in the context of technologies of surveillance, one's identity can change from frequent flyer, to security threat, to terrorist in a matter of (precisely, depending on the technology) calculated minutes that often hinge on the ethnic and national identity of that traveller.

Since ethnicity and national identity, as well as other bodily identities such as gender, ability, sexuality, or class, work in tandem with "hard" technologies to determine "good" from "bad" bodies, the book also approaches technology through the concept of somatechnics. Like geocorpographies, somatechnics highlights how the body is central to our experience of being in the world. The term conjoins *soma* (derived from the Greek and Latin word for "body") and technology to draw attention to the ways bodies find expression through social techniques of manipulation. We might think of these techniques as encompassing social conventions around dress and behaviour for men and women, or for religious and non-religious persons. Without these techniques, such bodies would not be identifiable as male or female, or as belonging to a particular religion. In this book, we show how these social techniques are complemented by hard technologies such as wearable health gadgets that measure fitness and wellbeing, geo-locative devices that determine how fast workers should complete their assigned tasks, and which bodies should be targeted for execution or exclusion from the social body as a whole. Viewed through the lens of somatechnics, the body is not a "natural" entity that *comes to be* changed by society but is *already* inscribed by the social training and techniques of presentation learned in the particular environment a body occupies.

In order to understand the global inequalities bought about by technologies of health and security, this book suggests that

- Geocorpography: how space affects and effects the treatment of bodies
- Somatechnics: how the body is manipulated and produced, and
- Biopower: how to maximise the health and security of the population

comprise a critical and politically informed framework through which to understand contemporary geopolitical debates and crises over new technologies as *extensions* of already existing divides between and within populations. Not all of the contributions engage with all three of these terms, but with their shared emphasis on the management of bodies and the role of racism in justifying and structuring these practices, an examination of a specific case study through one term will necessarily draw attention to the others.

The book is set out in three sections, marked by an extended engagement with the conceptual frameworks of Geocorpographies, Technologies, and Biopolitics. Throughout the sections, the themes delineated in the book's title are interrogated in the dimensions of space, mobility, and bodies. The first section of the book focuses on the corporeal and metaphorical bodies implicated in contemporary geopolitics; that is, the geocorpographic spaces produced by United States drone strikes that obliterate flesh and humanity with impunity, or the reach of the Enlightenment-era social body that supplants Australian Indigenous sovereignty with the sovereignty of the Crown. Contributions in this section also examine the geocorpographies produced by the global distribution of HIV-prevention drugs and the intensely surveilled space of the airport. The second section, 'Technologies', looks more intently at technological developments in digital surveillance and locative media in terms of the use of consumer lifestyle apps and the ecological and labour costs of producing avowedly "immaterial" technologies. In the final section of the book, contributors draw on Foucauldian notions of biopower and biopolitics to examine how policies and ideologies of security justify the valorisation of certain types of bodies and bodily practices at the expense of others. Here, contributors examine technologies of domestic drones, ageing, and the house as a way of delineating those bodies at risk of failure in reproducing a neoliberal consumer society.

In order to help readers navigate the different case studies, histories, and geographical spaces examined in the chapters, the remainder of this Introduction will explain in more detail the role of technology in population management in the work of Foucault. In particular, we will orient readers to his approach to power as a productive agent in *stimulating* action as opposed to its usual understanding as an oppressive force. We show how this approach to power is connected to his development of governmentality, which is used to elucidate the role of security in the governing of bodies and populations. We then discuss the three key terms mentioned above (Geocorpographies, Somatechnics, and Biopower) in more detail in relation to the specific contributions to the book, and in particular, how racism activitates the technological division and geographical targeting of particular bodies. Finally, we provide an overview of each of the individual chapters.

Technology, Power, Governmentality

Each of the essays contributed to this collection engages in some way or another with technologies of security, race, and biopower. In some of the contributions, these are what we might think of as social technologies, like the biopolitical discourse of ageing, the residue of sovereign power in settler states, or the symbolic institution of the 'house'. Other contributions deal with more material technologies, such as drones in both military and civilian contexts, pre-exposure HIV prevention drugs, and wearable health and fitness gadgetry. In examining technologies, we find ourselves examining strategies, relations, and flows (what Foucault would term capillaries) of power. Technologies enable and constrain the ability of power to impose itself in our lives. All technologies—that is, all socially-constructed and socially-realised material and cultural formations—have an impact on how embodied individuals experience their place in the field of power. In particular, the essays collected in this book describe and critique the ways in which contemporary technologies produce subjects according to their statistical risk or value in an atmosphere of generalised security, in relation to categories of race, and within a strategy of power centred on the body, through the dual poles of the individual and the population; that is, through biopower.

Power, thinking with Foucault, is not understood as something owned and exerted by the power*ful* over the power*less* but instead as a 'multiplicity of force relations' (1978, p. 92). In attempting to decode Foucault's four-part definition of power as multiplicity, process, support, and strategy (pp. 92–3), it is difficult to resist comparing power to the natural forces of the physical world, such as gravity. Although an analogy equating power with gravity is limited, in that it risks concealing the socially-contingent nature of power and its possibilities for resistance, it makes an apt comparison. Like gravity, power for Foucault has an inexorable and innate drive and direction. Like gravity, power is not measurable in a vacuum, but in the relations between objects that it links up and pulls apart. And, like gravity, power acts upon everything and emanates in some minute way *from* everything, often invisibly, to attach to all objects a vector and a velocity, a potential or actual magnitude. Foucault rejects understandings of power as 'a group of institutions and mechanisms', 'a mode of subjugation', or a 'system of domination exerted by one group over another'; although these may be the visible effects of power, the 'terminal forms power takes' (p. 92). Power for Foucault is not something owned or wielded, but the context and medium of relations between the objects under its purview.

This conceptualisation of power enables analyses of how ostensibly apolitical and non-state technologies function to reproduce dominant strategies of power. These strategies of power, as found in the essays collected here, tend to pull in certain directions, directions which might be considered normative (what is "normal"). They *gravitate* towards objectives of population, through discourses of security, race, and biopower. In this way, we argue that the essays contained in this book describe nodal points in the dispersed web of what Foucault terms 'governmentality'. Governmentality is the third major transformation of power described in Foucault's work, following sovereignty and discipline (though sovereignty and discipline do not vanish). It describes the constellation of objects, people, technologies, and apparatuses interconnected within a circulatory system of state and population, emphasising the way in which not only '*the* government' but also corporate and communal institutions perform the collective function of governing. Governmental power operates not through an inherent 'right' over life and death (as with the articulation of power under sovereign monarchies), or normative training in pursuit of docility (as in a disciplinary society), but through the ostensibly free play of interests and desires balanced against the security of the population. Foucault defines governmentality in three parts, the first (and most pertinent) of which is quoted here:

First, by 'governmentality' I understand the ensemble formed by institutions, procedures, analyses and reflections, calculations, and tactics that allow the exercise of this very specific, albeit very complex, power that has the population as its target, political economy as its major form of knowledge, and apparatuses of security as its essential technical instrument. (2009, p. 144)

When Foucault writes that governmentality has 'the population as its target', he means that government is an arrangement of objects in a complex relation of power that pulls in the direction of the overall health and wellbeing of the majority within the total population. That is, it contends with the question: how can government weigh the overall health of the population against the individual desires of citizens in such a way as to serve both? Government is therefore a form of numbers game which seeks balance in the 'protection of the collective interest against individual interests' (Foucault 2008, p. 65). Determining this balance (and correcting it where necessary, employing the disciplinary and biopolitical strategies of biopower) is the function of security, the 'principle of calculation' at the heart of liberalism. Freedom is the currency of this strategy of power, but it is a freedom that must be managed, 'constantly produced' (p. 65), and strained through the actuarial logic of security. The freedom of the collective population must not encroach the freedom of individuals, but the freedom of individuals must not impact the freedom of enterprise; 'the game of freedom and security is at the very heart of this new governmental reason' (p. 65).

The individual freedom to pursue desires and aspirations is an essential alternative to the generalised coercion that preceded in 'disciplinary societies' under monarchs and rulers who used force to directly discipline and punish the population. The impetus (cultivated by both state and non-state institutions of governmentality) to live a life that is at once productive, obedient, familial, healthy, consumerist, happy, career-driven, and inflected by countless other normative pressures drives self-regulation in contemporary Western societies with and through a degree of individual freedom that is both indispensable to late-capitalism and incompatible with exhaustively disciplinary or sovereign strategies of power. No longer punishable by kings, the liberal individual is free to pursue their interests. But, there is a risk to governing authorities that this freedom could be used against them or in a way that is counter to the perceived needs of the wellbeing of the whole population. The problem of modern government then is how to both nurture this freedom and direct it towards the goals of government. This directed, bounded, and manufactured individual freedom is the context for the chapters in this book, which each carve out a critical dialogue in the varied spaces of governmentality.

The essays in this collection then, tacitly or implicitly, describe technologies of security, race, and biopower which intersect in the web of apparatuses through which governmentalising power flows. They reflect the complex governmental entanglements of (in)security and capital; of the centrality of racism to expendable and mobile bodies; and of biopower, individual desire, and material technologies. Finally, they each emphasise the way power finds a 'surface of intervention' (Foucault 1978, p. 48) on the very bodies of its subjects, highlighting that the corporeal dimension does not escape power's grip in the passage of power from the discipline of sovereign rulers to the liberal freedom of government. The body and its capacities are still the key target of power for Foucault. The diverse collection of topics included here (the linear reader will soon leave behind the subject of visceral drone bombings for pharmaceutical representations in gay pornography) make innovative contributions to their respective fields, each employing a different disciplinary approach to a different subject and identifying emergent social and material technological formations of security, race, and biopower that invite further scholarly attention. While contributors to this collection examine a range of different technologies and locations, they cohere around analyses of the new possibilities for locating and managing bodies in space.

Geocorpography, Somatechnology, and Biopower

The geographies of space are a vital mechanism for the formation of bodies as subjects of power and the application of technologies to determine their utility. The mutual constitution of space and bodies as inherently geopolitical—and linked to war as a social ordering of bodies—is what led Pugliese to coin 'geocorpography' (2007). Geopolitics and geography refer here to an intimate and corporeal assemblage of the body and space rather than the traditional and abstract notions of citizenship and states as granting rights and freedoms through the possession of a national identity. Where public commentary and academic scholarship once extolled the virtues of digital media and technology in freeing people from physical locations, amounting to the 'death of geography' (see Sassen 2003), geocorpographies highlights how the corporeal and physical exigencies of labour and bodies in place are central to the technological collation and dissemination of data through drones, biometrics, medicine, and border security. As Tiziana Terranova has noted in the context of online cultures, 'Far from being an "unreal," empty space, the Internet is animated by cultural and technical labor through and through, a continuous production of value that is completely immanent to the flows of the network society at large' (2000, pp. 33–34).

In addition to materialising labour across geographical spaces, the contemporary technological capacity to manage bodies in geographical space has its antecedents in imperial and colonial forms of governance. As Jacinta Ruru notes, it was 'upon declaring the lived homes of Indigenous peoples "space" that colonial governments successfully overlaid their laws and rules on Indigenous place' (2008, p. 105). It is precisely because 'space' is an abstract entity to be managed that bodies, law, and histories can be vacated from its conceptual and practical applications. This colonial logic created a geocorpography that separated populations into peoples who were civilised with laws, who could manage and employ appropriate technologies to create 'places' and fill them with the appropriate bodies, while those "without" laws, who could not manage space or land, could be racially expelled and spatially constrained. Drawing attention to the colonial histories of racial and spatial formation highlights the continuities of contemporary forms of managing, securing, and dividing up space and the bodies that move, consume, and labour within it.

In order to understand the ways bodies are managed in space, each of the chapters explains how technologies are used to produce the body as a subject of power. This process of becoming a subject (or 'subjectification') through specific bodily techniques and technologies is understood in the book as somatechnical. Developed by researchers working in the then-Department of Critical and Cultural Studies at Macquarie University, somatechnics was conceived as a critical and metaphysical refusal to split the 'subject/ object' (Pugliese and Stryker 2009, p. 2) in accounts of the body's being in the world. Where traditional philosophical accounts of the body view it as a tool to be used and directed by the mind (as an object or thing to be controlled and overcome), somatechnics suggests that our bodies and their capacity for transformation affect our sense of self as a subject. And once we are made subjects, we become subject to power relations. A common approach to the body's role in life is to see it as a machinic extension of thought. For instance, *I* cut myself on paper so *I'm* fastening a band-aid around *my* finger. A somatechnical perspective would invite a reconsideration of the inter-locking objects and bodies producing unexpected collisions: the paper slips from hands that are moisturised, soft tissue hitting sharp edge, blood surfaces, a band-aid's adhesive secures itself to the matching palette of skin. In this somatechnical scene, slippery hands are productive of a gendered self, the office setting a professional occupation, the colour correspondence of plastic tourniquet to flesh signals whiteness as the default corporeal phenotype. Such identities are simultaneously registered and bought forth by the bodily techniques of modification that are social, spatial, and technological. These techniques are implicated in power relations at both an individual level (in terms of the desire for gendered expression or occupation) and a broader population level (in terms of which labouring bodies are deemed "normal" and "useful"). One aim of the somatechnic lens is to challenge and problematise notions of the body as a natural and neutral extension of the self.

Considered as somatechnical, the body does not become a technology through representation or the external utility of tools and objects but is always-already figured as a technological site for the historical, spatial, and political embodiment of power. Connecting *techne* with *soma* shows how 'we have never existed except in relation to the *techne* of symbolic manipulation, divisions of labor, means of shelter and sustenance, and so forth' (2009, p. 2). This 'so forth' gestures towards both the extraordinary and quotidian manifestations of the body's abilities to at once exercise power over life and remain subject to spatial and political relations of dominance.

Joseph Pugliese and Susan Stryker situate somatechnics as the bridge between Foucault's designation of the 'anatamo-political' exercise of power through individual bodies, which functions at a disciplinary level to produce them as subjects, and the 'biopolitical' that centres on the 'mechanisms of life' as a whole (2009, p. 3). As explained above, liberal forms of governance require subjects who are free to act but direct their actions towards the ends of governmentality. In doing so, subjects are both made aware of themselves as individuals (with the individual freedom to self-discipline and make themselves "normal") and as constituents of a larger population (whose overall health must be secured for the "good" of all). These dual strategies (the anatamo-political and the biopolitical) constitute biopower. Somatechnics explains how the body is implicated in systems of power which both produce a docile subjectivity that can be disciplined by bodily techniques and the larger institutional ordering and management of bodies that takes place at the level of populations. Understood through somatechnics, the chapters in this book illustrate 'the role of the body in the production of knowledge' (Pugliese and Stryker 2009, p. 1) regarding the modes of identification (as gendered, as raced, etc.) through which power takes hold of the corporeal.

Critically, somatechnics draws attention to the capillarising and activating role of racism in Foucault's genealogy of power. For Foucault, colonisation provides the impetus for a biological form of racism that spatially segregates global populations and justifies the exclusion or killing of aleatory bodies within the population (2003, p. 257). He argues, 'In a normalizing society, race or racism is the precondition that makes killing acceptable' and 'Once the State functions in the biopower mode, racism alone can justify the murderous function of the State' (p. 256). Racism is understood by Foucault as encompassing biological traits designated as unproductive or undesirable for the state, which can also include ability or sexuality. He qualifies his use of the term 'killing' by explaining,

> When I say 'killing', I obviously do not mean simply murder as such, but also every indirect form of murder: the fact of exposing someone to death, increasing the risk of death for some people, or, quite simply, political death, expulsion, rejection, and so on. (p. 256)

As a result of the state's role in preserving life that is productive and necessary for the social body, war becomes an organising rationale for and truth-effect of the arrangement of bodies within a population, stimulated by 'the theme of racism' (p. 257). Subjects are encouraged to see themselves in a war over the resources that secure life, which therefore mitigates the potential for a critique or protest against the governing authorities that manage these resources.

Walter Mignolo describes biopolitics and biopower as 'regional critical concepts' since they explain how racism is inscribed in the development of modern European states (2012). The biological construction of race instantiated asymmetries of bodily value in the context of colonial conquest, where war was a direct exercise rather than an operation of governance. Beginning in the seventeenth century, the epistemological and ontological construction of non-Europeans as 'deficient' supplied the rationale for global capitalism; 'you cannot exploit and expropriate an equal' (2012). Thus the racist 'logic of coloniality' created and justified the instrumentalisation of 'deficient' bodies for slave labour and the production of surplus goods. Such an economy 'introduced … the dispensability of human life' (2012) that animates contemporary global hierarchies of capitalism and their attendant geopolitical inequalities.

Understood as a strategy of governance that intersects with the production of expendable life, all of the contributions in the book touch on the 'somatechnological instrumentalization of racism' (Pugliese and Stryker 2009, p. 4) as a means of calculating the body's utility. In each of the contributions, the affected bodies employ and are employed by technologies to become aware of themselves as subjects: as belonging to identities that are biologically and economically productive, defective, or simply surplus to a capitalist regime within a particular spatial and political locale.

Geocorpography, somatechnology, and biopower structure the volume's engagement with the racialised spatial ordering of bodies and the technological formation of subjects worth preserving or destroying, worth making live or letting die (Foucault 2003, p. 247): geocorpographies which inscribe in the body the violent machinations of geopolitics founded on empire and colonisation; somatechnologies which always-already produce the body of the subject and cannot be extracted from it; and biopower which from one pole disciplines and normalises bodies, and from the other pole, identifies, sorts, manipulates, and exterminates them.

The Content and Data of Bodies

All contributed chapters are situated in the space of geocorpography, somatechnics, and biopower. From a Foucauldian standpoint, the contributions outline how bodies are always technologies of a kind of "security", always reproducing or articulating the objectives of governmental power, of the phenomenon called the "population", not in spite of but because their corporeal vulnerability (their capacity for failure, their capacity for revolt) is rendered as risk within these paradigms of power.

In Section One: Geocorpographies, the first chapter, by Joseph Pugliese, develops 'bioinformationalisation of life' as a means of accounting for the 'the interlocking of the [National Security Agency] NSA's metadata with the [Department of Defense] DoD's algorithmic formulae used to conduct drone kills in which often the identities of those killed are not known' (p. 4). Disrupting the celebratory narratives of technological progress that posit the scientific accuracy of drone strikes, and therefore "humane" operation of war, Pugliese recounts the horrifying violence unleashed through drone warfare because its inexact targeting is predicated on a metonymical association between the Arab body of one "terrorist" for another. Simply occupying a geographical space and identity is enough to render all bodies suspect in this war. Drone warfare is enacted through cell phone use where the consumer's own carrier signals can be locked onto nearby towers to relay geolocative telecommunication data to drone operators. In this way, the use of phone technology somatechnically instrumentalises a terrorist or "terrorist-like" body ready for killing by harnessing the subject's own consumption practices against them. Locking drone targets to the carrier data of moving bodies also constructs 'a bounded locus' for an otherwise fluid and mobile 'transnational terrorism' (Gregory 2004, p. 50). The chapter quotes NSA General Counsel Stewart Baker's overview of this process: 'metadata absolutely tells you everything about somebody's life. If you have enough metadata, you don't really need content' (p. 4). This reduction of the biography of those who carry phones to superfluous 'content' is illustrative of what Derek Gregory describes as 'Modern cartographic reason' wherein 'electronic, mediatized extensions' of spatial warfare 'relies on … high-level, disembodied abstractions to produce the illusion of an authorizing master-subject' (2004, p. 54). That up to 90 per cent of drone strikes result in casualties of innocent civilians (p. 7) is not surprising given the confident abstracted techno-logic employed by the DoD. Pugliese contends these casualties cannot be labelled 'errors' or 'mistakes' when the extrajudicial drone killings are 'predicated on anthro-pocentric hierarchies of life' (p. 8) which biopolitically designate some bodies as justified objects of slaughter in the United States' right to self-defence.

Pugliese's chapter introduces two critical themes of the volume: the role of empire and state sovereignty in determining what resources and technologies can be used to preserve life and the entanglement of technological consumption in the spatial ordering, emaciation, and targeting of "deserving" and "undeserving" bodies. The mediation of which bodies, in which spaces, are deserving of pleasure, life, and security is explored by Joshua Pocius in Chap. 2. Pocius expands on Pugliese's work to focus on the biopolitical and geocorpographic role of antiretroviral medication. The chapter situates 'the emergence of pre-exposure prophylaxis (PrEP) as a biomedical technology which straddles discourses of public health and erotic pleasure' (p. 22). Developing the term 'eRotics', Pocius shows how there is a connection between preservation and pleasure in medical technologies of risk mitigation. For gay male bodies in the Global North, PrEP drugs potentially facilitate new ways of exercising identity and fraternity that challenge older forms of community built out of the AIDS (acquired immune deficiency syndrome) crisis (and generate new forms of pornography and erotic consumption). For bodies in the Global South, however, prevention and treatment remains sluggish due to geopolitical and medical assumptions that the drug will be consumed irresponsibly. These geocorpographies of consumption and risk mitigation are built into the mapping of the virus where the 'strain' most predominantly found in Western countries is given comparatively more attention and treatment than the virus' manifestations in non-Western spaces (p. 35). Pocius posits that the possibilities for sexual practices such as barebacking to be enabled by PrEP and imagined through pornography must be situated in a geocorpographic continuum with the bodies in the Global South exposed to bare life without pharmaceutical protection.

Where Pocius' chapter focuses on the risks of PrEP use by individuals, weighed against the security and health of the social body as whole, Holly Randell-Moon considers the historical and contemporary role of the British monarchy as likewise preserving the security and durability of the social body on behalf of Commonwealth nations. Arguing that bodies (both symbolic and literal) were central to the "discovery" and possession of new territories during colonialism, Randell-Moon shows how a geocorpographic configuration of Commonwealth nations was and still is essential for the white, Protestant Crown to claim sovereign authority in settler states like Australia. Crucially, the apparent secularity of law in settler states operates as 'juridical affirmation of settler sovereignty and the extinguishment of Indigenous sovereignties as competing forms of authority' (p. 42). Randell-Moon demonstrates with the example of succession laws in the United Kingdom and their patriation into the domestic secular laws of Commonwealth nations that secular law does not function as a demonstration of the autonomy of settler states to make their own laws. Rather, the secular law of settler states serves to capillarise and activate a British sovereign power that is theocratic in its ties to the Church of England and the British monarchy. She focuses on the UK Succession to the Crown Bill 2012, which was fast-tracked in order to prepare for the possibility of a female heir born to the Duchess of Cambridge, as it was feared that the potential ascension instead of a younger male sibling (without cause, it later proved) would have sparked neoliberal outrage. The succession amendments that followed Britain's example throughout the Commonwealth, despite being framed through discourses of liberal equality, 'work to abstract the bodies of the Monarchy from the geopolitical and imperial histories of war, discovery, and settlement' (p. 56). The chapter concludes that settler "secularity" is primarily a secularity from Indigenous sovereignty and sovereignties which would challenge the Crown possession that underpins secular law.

Sunshine Kamaloni's conclusion to the Geocorpographies section is a first-person exploration of the geocorpographies of borders, bodies, and security. Critically reflecting on the experience of being a black woman in the heavily-regimented space of an airport, the chapter's title, 'What are you doing here?' is a declarative assertion of privilege. The title's address to the author provokes feelings of uncertainty, implicit exclusion, and scrutiny which characterise her every visit to the airport. As well as recounting specific experiences, Kamaloni examines the space of the airport as a border, a space for surveillance, a post-9/11 nexus of geopolitical insecurities, and a racialised and technologised geocorpography. She describes in a confessional tone how passing through Melbourne's Tullamarine airport makes her 'acutely aware and self-conscious of my body and the otherness attached to it' (p. 64), offering something that is not present in any other contribution to this book: an affective experiential account of geocorpographic violence. In mobilising this self-ethnography, the chapter's methodology is in line with Michael Shapiro's notion of the 'violent cartographies' constituted through the ontological placement of self and other (1997). Kamaloni recalls two particular incidents in her narrative: a near-collision with a middle-aged white woman, wherein the corporeal borders of differently-marked and labelled bodies produces discomfort and shame, and her surprise at a customs officer mistakenly grouping her with a stranger, due to the matching colour of their skin. Throughout this chapter, the spectre of borders persists: national borders, airports as borders 'within borders', bodily borders, and biopolitical borders as racialised caesurae. The airport, Kamaloni finds, is a space striated with immeasurable and invisible borders, of which her blackness makes her poignantly aware.

If the technologies of security at the airport serve a larger biopolitical purpose of risk management within the social body, by making certain travellers aware of themselves as subjects of risk, Chaps. 5 and 6 focus on a differently subjectified body, but one nonetheless produced by geocorpographic and biopolitical currents: the consumer body. In the first chapter of Section Two: Technologies, Ryan Tippet examines the 'Internet. org' project by Facebook to offer free internet access in under-developed regions of the world, seeking to problematise the campaign's charitable and revolutionary rhetoric through a framework of 'corporate geocorpography'. Whereas Pugliese's original concept emphasised the corporeal violence of state geopolitics, Tippet demonstrates with corporate geocorpography how the business campaigns of companies like Facebook can produce parallel spaces of '*seductive and inclusive enmeshment of the flesh and blood of the body within the economic geography of race, technology, and imperialism*' (p. 82). In centring the body to these geographies of corporate expansion, Tippet seeks to demystify the 'immaterial' rhetoric of the 'global knowledge economy', using surveillance theory and Mark Andrejevic's concept of digital enclosure to anchor Internet.org's mobile app and public relations output to the racialised and labouring bodies of its subjects. Such an undertaking is in line with Saskia Sassen's contention that 'There is today no fully virtualized firm or economic sector' (2003, p. 22) and 'Information technologies have not eliminated the importance of massive concentrations of material resources' (p. 22). Ultimately, and in spite of its apparent altruism, Internet.org cannot be disentangled from the economic motivations of its corporate members, and the project 'to bring affordable internet access to the two thirds of the world without it' (p. 81) fits neatly within a surveillance and corporate geocorpographic critique wherein the ethos of 'The more we connect, the better it gets' (cited in p. 93) veils an exploitative and racialised imperial project of commodification in emerging digital enclosures.

In Chap. 6, Brett Nicholls situates the rise of wearable technologies of self-quantification (and their companion mobile apps) in the critical framework of Deleuze's 'societies of control'. Nicholls analyses the discourse of 'self knowledge through numbers' (cited in p. 102) by asking what the personalised 'reconfiguration of bodies, technology, and data' (p. 102) does for and to the self-responsible and health-conscientious modern liberal subject. Nicholls deploys Marxian and Foucauldian theories to support his critique of the technologies of 'Everyday Modulation', elaborating the densely-collaged social context that preconditioned the emergence of the tangible and discursive technologies of self-quantification. Like Tippet in Chap. 5, Nicholls finds that Foucault's 'disciplinary societies' provide an insufficient framework to describe the forms of surveillance enabled by mobile quantification apps, whereas Deleuze's control societies account for the always-on, automatic, every-day nature of wearable health motivation technologies. These are technologies which instrumentalise bodies in the name of security—in the governmental sense of the security of healthy and productive bodies.

While Tippet and Nicholls examine how consumers of information communications technology and consumer electronics (CE) are culti-vated, Sy Taffel's concluding chapter in the Technologies section focuses on the production processes of the devices themselves. Taffel asks us to 're-think the material politics of digital culture' (p. 122) by examining the toxic waste produced by the extraction of rare earth elements, their assembly, use, and recycling in CE. Scholars have approached the life-cycle of CE and the proliferation of screens through media ecology (see Deuze 2012) and the transformations of cognitive capabilities required to labour and participate in the attention economy (see Stiegler 2012). Taffel's work shows how our media life is inextricably tied to the life-cycle and biological transformation of bodies in the Global South. Though their labour is (often fatally) essential to the proliferation of CE, the role of these bodies is invisibilised in the production chain of smart technolo-gies. When these technologies are exported to consumers in the Global North, large corporate chains utilise them to implement "smart" work practices that calculate to the minute the optimal completion times for industrial tasks. These financial pressures on the extraction, assembling, and implementation of CE exemplify the shifting geopolitical construc-tions of consumer privilege which centre China as a global economic power. Both older and newer forms of imperial and colonial geographies have effected the contemporary capitalist flows of digital technologies, and their production exigencies intersect with racialised and classed con-figurations of disposable bodies across Asia, the Global South, and the West. Caught up in the cycle of extraction, production, and utility of

CE, precarious workers across the globe are differently enmeshed in the capitalist gears of surplus profit generated through technical efficiency. To document the environmental and somatic toxicities of CE is to then corporealise dominant conceptions of the knowledge economy as "weightless" and "green".

In the final section of the book, Biopolitics, Caitlin Overington and Thao Phan's focus on consumer drones illustrates another geocorpographic consumption cycle where technologies associated in one spatial context with death and destruction are transposed into pleasure and indulgence in another. Here the book revisits drones and examines their resignification from killing apparatuses to banal hobby and feature of urban life. In order to map this transition, Overington and Phan draw on Hannah Arendt's notion of banalisation, showing how technologies of war come to fit comfortably within existing security apparatuses such as Closed-Circuit Television. In this re-purposing of drones, and following Michael Hardt and Antonio Negri's concept of 'Empire', war becomes a 'general phenomenon, global and interminable' (cited in p. 148). Overington and Phao find, therefore, 'that the civilian uptake of drones is demonstrative of the banalisation of war within this regime' (p. 148). Drones as hobby exemplify the role of conflict and war in the global distribution and flow of technology and capital into city spaces. Being able to access and use drone technology for leisure implicates drone consumers within biopolitical 'ways of seeing' and broader security regimes of generalised surveillance.

Whereas Overington and Phan identify the blurring of an urban/ war and hobby/ weapon set of surveillance binaries, Chap. 9 examines the biopolitical apparatus enclosing individuals on the wrong side of an age divide. David-Jack Fletcher situates his analysis in the context of biogerontology, where ageing bodies are constructed 'as "diseased" within a biomedical framework' (p. 168). Drawing on the work of Giorgio Agamben and Foucault, Fletcher argues that aged bodies are categorised as unproductive and therefore quarantined from normal society under this biomedical paradigm. Crucially, this occurs through a somatechnics of freedom whereby the elderly are positioned as desirous of this exclusion for their own good. For Melinda Cooper, biotechnologies and commercialised life sciences emerge from a post-industrial valorisation of life as a surplus (2015). Fletcher locates anti-ageing medicine, 'emerging telomere and stem cell based therapies' (page), as part of an older humanist desire to overcome mortality itself. Evoking Foucault's anatamo-politics, Fletcher contends that biogerontology depends on 'the conception of the individual body as a machine' (p. 172). From the perspective of 'racist' biopower, when age intervenes in the efficiency of that machine, the killing or removal of the deteriorating individual body is a justified act of war to preserve the social body.

Jillian Kramer continues the discussion of war as a biopolitical form of social organisation in the book's final chapter, with an analysis of the settler colonial machinations of the Northern Territory Emergency Response (NTER) in Australia. Colloquially known as the "Intervention", the NTER was initiated in 2007 by the then John Howard-led Coalition government as a response to reports of child sexual abuse in remote Northern Territory Aboriginal communities. The response included alcohol restrictions, compulsory child health checks, quarantining of social security payments, the abolishment of the permit system (where visits to and residence on communal land were undertaken with permission from the local Aboriginal community), a military presence to enforce these measures, and controversially, the suspension of the *Racial Discrimination Act 1975* (Cth) so that the measures could be racially discriminatory and apply to all Indigenous residents in the targeted communities. Kramer's point of entry is the symbolic role of the 'house' in Intervention policy and discourse as both an institution of the settler-colonial state and a site of Indigenous resistance. Couched in militaristic rhetoric and appeals to the logic of security, the Intervention sought to displace the agency of Indigenous communities through land reacquisition, punitive measures, and 'economic regimes', citing the inability of Aboriginal people to self-govern in accordance with Australian settler-state governmentality. Drawing from Paula Chakravartty and Denise Ferreira da Silva, Kramer shows that the figure of the house mobilised by the settler-colonial state must be conceptualised as a thoroughly racialised construct (Chapter 10), especially as it functions in the Intervention as an indexical signifier of 'coextensive notions of "private property", "economic security", and "business management," to (re)produce the legal fiction of *terra nullius*' (p. 191). Kramer's contribution demonstrates the way in which 'normativity', despite functioning as an incentive for self-discipline under the freedom-oriented strategy of governmental power, is still a restricted and racialised white capitalist normal, leaving 'aleatory' Indigenous communities subject to figurative 'killing' by the disciplinary and biopolitical tactics of biopower.

United under the themes of geocorpographies, technologies, and biopolitics, the contributions in this edited collection disclose how the connections between space, race, and bodies are central to the emergence and refinement of technologies of surveillance, medicine, law, and war. Contested in every chapter are relations of power structured by multiple competing technologies of sovereignty, discipline, and government, and our focus in assembling these chapters under the heading *Security, Race, Biopower* has been to demonstrate the way these technologies—no matter how weightless and "ethereal", and regardless of whether they operate on a mass or individual level—find a surface of intervention on corporeal bodies.

References

Cooper, M. (2015). *Life as Surplus: Biotechnology and Capitalism in the Neoliberal Era.* Seattle: University of Washington Press.

Deuze, M. (2012). *Media life.* Malden: Polity Press.

Foucault, M. (1978). *The history of sexuality Vol. 1.* New York: Pantheon Books.

Foucault, M. (2003). *"Society must be defended": Lectures at the Collège de France, 1975–1976* (D. Macey, Trans., & M. Bertani & A. Fontana, Ed.). London: The Penguin Press.

Foucault, M. (2008). *The birth of biopolitics: Lectures at the Collège de France, 1978–1979* (G. Burchell, Trans., & M. Senellart, Ed.). New York: Palgrave Macmillan.

Foucault, M. (2009). *Security, territory, population: Lectures at the Collége de France, 1977–1978* (G. Burchell, Trans., & M. Senellart, Ed.). New York: Palgrave Macmillan.

Gregory, D. (2004). *The colonial present.* Oxford: Blackwell Publishing.

Mignolo, W. (2012, September 22). *The prospect of harmony and the decolonial view of the world.* http://waltermignolo.com/the-prospect-of-harmony-and-the-decolonial-view-of-the-world/. Accessed 23 Oct 2015

Pugliese, J. (2007). Geocorpographies of torture. *Critical Race and Whiteness Studies*, 3(1). http://www.acrawsa.org.au/. Accessed 23 Sept 2014.

Pugliese, J. & Stryker, S. (2009). The somatechnics of race and whiteness. *Social Semiotics*, 19(1), 1–8.

Ruru, J. (2008). A maori right to own and manage national parks? *Journal of South Pacific Law*, 12(1), 105–110.

Sassen, S. (2003). Reading the city in a global digital age: Between topographic representation and spatialized power projects. In L. Krause & P. Petro (Eds.), *Global cities: Cinema, architecture and urbanism in the digital age*. New York: Rutgers University Press.

Shapiro, M. (1997). *Violent cartographies: Mapping cultures of war*. Minneapolis: University of Minnesota Press.

Stiegler, B. (2012). Relational ecology and the digital pharmakon. *Culture Machine*, 13, 1–19. http://www.culturemachine.net/index.php/cm/article/view/464/501. Accessed 8 Oct 2015.

Terranova, T. (2000). Free labor: Producing culture for the digital economy. *Social Text*, 18(2), 33–58.

Acknowledgements

This book attempts to draw attention to the ongoing struggles over geography and the security enacted to preserve territorial integrity. As such, the editors offer a small gesture to the continuing contestations of state security and biopower by acknowledging mana tangata whenua and the lands known as Ōtepoti where this book and the spatial injustices it names was assembled.

Security, Race, Biopower emerged from the 2014 *Space, Race, Bodies: Geocorpographies of City, Nation, Empire* conference held at the University of Otago, December 8–10th.[1] The conference was funded by a Humanities Research Grant from the University of Otago and we are grateful to the Division, along with the Sexuality Research Group and the Postcolonial Studies Research Network at Otago for their support of the event. The Somatechnics Research Network, hosted by the Institute for LGBT Studies at the University of Arizona, along with Susan Stryker, also provided crucial support. Without the generosity of these centres of research, this book would not have come to fruition. Sincere thanks also go to the conference committee, Mahdis Azarmandi, Katharine Legun, Alex Thong, and Maud Ceuterick, for their friendship, creativity, and commitment to the conference and its research outcomes.[2] Chris Prentice and Brett Nicholls also provided ongoing support for the conference and book project; their generous collegiality is much appreciated.

We are sincerely thankful for the guidance and encouragement from Vijay Devadas whose critical insights immeasurably strengthened our own contributions and the book overall.

We also owe thanks to the anonymous reviewers, who generously offered their time to provide critical feedback and essential advice on each chapter.

Joseph Pugliese's rigorous and sustained attention to social justice and the multifarious ways law intervenes, divides, and touches bodies served as the inspiration for the collection, and we are grateful for his contribution to the book and the opportunity to explore and expand his ideas. We also thank the contributors to the book for their wonderful chapters and the opportunity to create new scholarly and political collaborations.

Holly would like to thank her parents, her brother Arthur, and family for their love and support during her academic adventures.

Ryan thanks his family and friends for their support, and especially Ellena for her encouragement, belief, and advice.

Finally, our thanks to Holly Tyler and the publishing team at Palgrave for their initiation of this project, along with their brilliant editorial support and assistance.

Notes

1. For more information on the conference and the *Space, Race, Bodies* research collective, please visit our website: http://www.spaceracebodies.com.
2. The journal companion to this book is the special issue of *Somatechnics* (vol. 6, no. 1) entitled, "Geocorpographies of Commemoration, Repression and Resistance", which is edited by Mahdis Azarmandi, Elaine Carbonell Laforteza, and Maud Ceuterick.

Contents

Notes on Contributors

David-Jack Fletcher is a PhD candidate at Macquarie University, Australia. His research focuses on the medicalisation of age through both hard and soft somatechnologies. His thesis argues that the development and deployment of anti-aging somatechnologies challenges notions of the human.

Sunshine Kamaloni has a PhD in communication and cultural studies from Monash University, Australia. Her research interrogates the mechanisms that sustain racism in the twenty-first century with a particular focus on the intersection of spatial practices, racialisation, corporeality, and experiential reflection. She is interested in how people experience and articulate their everyday experiences of race.

Jillian Kramer is a PhD candidate and teaching fellow in cultural studies at Macquarie University, Australia. Her thesis uses critical race and whiteness and critical legal studies frameworks in order to examine the Northern Territory Intervention. The thesis aims to explore the foundational role that race, specifically whiteness, plays in reproducing the settler-colonial order.

Brett Nicholls works at the University of Otago, New Zealand, in the Department of Media, Film and Communication. He has published work on postcoloniality, most recently with a special journal issue on the work of Edward Said, and he has published work on technology and media, specifically on video games.

Caitlin Overington is a PhD candidate in the School of Social and Political Sciences at the University of Melbourne, Australia. Her research explores CCTV and surveillance through the themes of banality, visibility, and gender. She is a member of the Research Unit in Public Cultures Graduate Academy at the University of Melbourne and the Urban Environments Network.

Thao Phan is a PhD candidate in the Media and Communications program at the University of Melbourne, Australia. Her research interests are in feminist technoscience, and her dissertation addresses the gendered dimensions of Artificial Intelligence discourse. She is a postgraduate affiliate of the University of Melbourne's Transformative Technologies Research Unit.

Joshua Pocius is a PhD candidate in the School of Culture and Communication at the University of Melbourne, Australia. His dissertation explores the temporal, spatial and biopolitical aspects of screen-mediated representations of HIV/ AIDS in the "post-AIDS" era. He is a member of Melbourne University's Research Unit in Public Cultures Graduate Academy.

Joseph Pugliese is Research Director of the Department of Media, Music, Communication and Cultural Studies, Macquarie University, Australia. His book, *State Violence and the Execution of Law: Biopolitical Caesurae of Torture, Black Sites, Drones* (Routledge, 2013), was nominated for the Hart Socio-Legal Book Prize 2013 and the Law and Society Association Herbert Jacob Book Prize 2013.

Sy Taffel is a lecturer in media studies at Massey University, Aotearoa New Zealand. He has published work on the political ecologies of digital media, media and materiality, media and activism, and pervasive/locative media. He is a co-editor of *Ecological Entanglements in the Anthropocene* (Lexington, 2017).

Randell-Moon is a lecturer in communication and media at the University of Otago, New Zealand. She has published on race, religion, and secularism in the journals *Critical Race and Whiteness Studies*, *Borderlands*, and *Social Semiotics* and in the edited book collections *Mediating Faiths* (2010) and *Religion After Secularization in Australia* (2015).

Ryan Tippet is a PhD candidate at the University of Otago, New Zealand. His research focuses on surveillance and social media, looking in particular at the constitutive relationship between the two, while his previous work has examined surveillance and security discourses in reality television.

Part I

Geocorpographies

To think about bodies—their capacities, vulnerabilities, corporealities—is to think about our experience of the world as grounded in bodily being. On this ground, our bodies are relational because they connect and collide inter-corporeally with other bodies, objects, and environments. Developed as a way of explaining how bodies are 'geopolitically situated' through 'regimes of visuality' (2007, p. 12), Joseph Pugliese's notion of geocorpographies seeks to challenge the abstract logic that would render geography, politics, and movement across borders seemingly devoid of embodied experience and the processes that inscribe bodies with meaning and identity. The four chapters in this section examine the geographies of security and law as corporealised in drone warfare, medicine and pornography, the British monarchy, and airport surveillance.

Pugliese's own chapter opens the book with a piercing account of the United States' drone kill program in Pakistan and Yemen. How this program renders life killable is made plain in the stark jargon of military officials who use words like 'data', 'content', and 'targets' to describe these killings and lend scientific accuracy to the bloodshed unleashed by drone warfare. Pugliese counters this dispassionate scientism by coining the term *bioinformationalisation of life* to reintroduce the body and life back into the de-humanising, data-driven language that frames the deployment of drone technologies. Joshua Pocius' chapter continues the focus on the transformation of bodies through scientific technologies in the context of antiretroviral medicines in the prevention of human immunodeficiency virus/ acquired immune deficiency syndrome (HIV/AIDS). The role of bodies, and the right kinds of bodies, in preserving Crown rule is the subject of Holly Randell-Moon's Chap. 3. The section concludes with Sunshine Kamaloni's experiential account of airports and racial profiling. In describing the quotidian encounters that take place between bodies in the airport, Kamaloni articulates how race underprops somatic visibility such that non-white bodies are isolated for being 'out-of-place' and hence subject to surveillance.

In their focus on the geographies of security and preservation that designate lives worth saving or letting die, the chapters engage differently with geocorporgraphies as critical praxis to account for the politics of embodied experience and spatial ordering. Pugliese considers the targets of drone kills fatally coextensive with their geocorpographies because their spatial location is read as evidence of their "suspect", and hence killable, status. He extends the term to consider the *geobiomorphology* that occurs when the carnage of drone warfare renders the corporeality of humans, animals, and other objects, indistinguishable. Kamaloni likewise situates her own body as generative of and inscribed by geocorpograhies of race, space, and place. Pocius meanwhile, tracks the asymmetry of resources tied to the HIV/AIDS pandemic as determined by geocorpographies which also work to simultaneously render "viral" bodies a geocorpography. Randell-Moon suggests that Crown rule functions geocorpographically to situate particular bodies (Anglican, English) as necessary for the preservation of settler state sovereignty, which functions to obscure already existing Indigenous sovereign bodies and laws. The chapters each reveal, as per Pugliese's original conception, that attempts to expunge the body—both literally and epistemologically—from its geographical location are saturated in the violence and power relations of race.

1

Death by Metadata: The Bioinformationalisation of Life and the Transliteration of Algorithms to Flesh

Joseph Pugliese

What are the effects of the increasing use of surveillance technologies in the conduct of contemporary wars? What are the relations of knowledge/power that undergird instrumentalising technologies that focalise life, transmute it into generic and anonymous data, and thereby render it killable through lethal militarised targeting?[1] Drawing on the revelations of Edward Snowden and two former drone operators, in this chapter I attempt to answer these questions by pursuing the lines of convergence between the United States' Department of Defense (DoD) and the National Security Agency (NSA) in the conduct of the US' drone kill program. In the first part of the chapter, I focus on new tracking

I wish to thank Holly Randell-Moon for her invitation to contribute the keynote presentation, upon which this chapter is based, at the *Space, Race, Bodies: Geocorpographies of City, Nation, Empire* conference, 8–10 December 2014, Otago University, Dunedin, New Zealand. I am profoundly grateful to Constance Owen for her brilliant and enduring research assistance.

J. Pugliese (✉)
Department of Media, Music, Communication and Cultural Studies, Macquarie University, Sydney, NSW, Australia

© The Author(s) 2016   **3**
H. Randell-Moon, R. Tippet (eds.), *Security, Race, Biopower*,
DOI 10.1057/978-1-137-55408-6_1

technologies developed by the NSA that have been incorporated into the DoD's drone targeting program. Specifically, I examine the interlocking of the NSA's metadata with the DoD's algorithmic formulae used to conduct drone kills in which often the identities of those killed are not known. In the latter part of the chapter, I situate what I will term the *bioinformationalisation of life* within the geocorpographies of Pakistan and Yemen in order to disclose the violent transliteration of abstract metadata to flesh.

Death by Metadata and the Bioinformationalisation of Life

In the course of a debate at Johns Hopkins University on the topic of the NSA's bulk surveillance programs, Michael Hayden, former Central Intelligence Agency (CIA) and NSA director, confirmed NSA General Counsel Stewart Baker's observation that 'metadata absolutely tells you everything about somebody's life. If you have enough metadata, you don't really need content' (cited in Cole 2014). I want, presently, to discuss this notion of metadata as superseding the need for content but, at this juncture, I want to focus on Hayden's comments. After remarking that Baker's observation was 'absolutely correct', Hayden asserted: 'We kill people based on metadata' (cited in Cole 2014). As has been well documented, US drone operators rely on metadata in order to determine what targets to terminate on their kill lists. Furthermore, as was evidenced by documents released by Edward Snowden, 'the agency analyzes metadata as well as mobile-tracking technology to determine targets, without employing human intelligence to confirm a suspect's identity' (RT 2014). An unnamed drone operator succinctly outlines this practice: 'People get hung up that there's a targeted list of people … It's really like we're targeting a cell phone. We're not going after people—we're going after their phones, in the hopes that the person on the other end of that missile is the bad guy' (cited in RT 2014). Two things are operative in these collected remarks that are worth unpacking: that if you gather enough metadata, it will supplant the need for 'content'; and that human targets, in the context of metadata-driven drone kills, become so somatechnically instrumentalised as to be entirely coextensive with the technology they use—in this case, their phones. This practice is further evidenced by a former drone operator who worked with Joint Special Operations Command. The former drone operator has disclosed the expansive and mobile dimensions of the NSA's surveillance and tracking sweep:

> the NSA doesn't just locate the cell phones of terror suspects by intercepting communications from cell phone towers and Internet service providers. The agency also equips drones and other aircraft with devices known as 'virtual base-tower transceivers'—creating, in effect, a fake cell phone tower that can force a targeted person's device to lock onto the NSA's receiver without their knowledge. That, in turn, allows the military to track the cell phone to within 30 feet of its actual location, feeding the real-time data to teams of drone operators who conduct missile strikes or facilitate night raids. (cited in Scahill and Greenwald 2014)

The NSA's program deploys 'advanced mathematics to develop a new geolocation algorithm intended for operational use on unmanned aerial vehicles (UAV) flights' (Scahill and Greenwald 2014). The former drone operator has also revealed crucial details about the NSA's use of the tracking program Geo Cell. Geo Cell identifies and 'geolocates' a tracked cell phone or SIM card without necessarily being able to determine who the person on the other end of the phone is:

> 'Once the bomb lands or a night raid happens, you know that the phone is there,' he says. 'But we don't know who's behind it, who's holding it. It's of course assumed that the phone belongs to a human being who is nefarious and considered an unlawful enemy combatant ... They might have been terrorists,' he says. 'Or they could have been family members who have nothing to do with the target's activities ... It's really like we're targeting a cell phone'. (cited in Scahill and Greenwald 2014)

The practice of killing by metadata underscores the intensification of what I will term the *bioinformationalisation of life*. Drawing upon a Heideggerian critique of contemporary science, the bioinformationalisation of life results from positivist science's demand 'that nature reports itself in some way or other that is identifiable through calculation and that it remains orderable as a system of information' (Heidegger 1978, p. 304). The convergence of metadata systems and digitised identification systems exemplifies the rendering of life into an orderable system of information through the application of algorithmic formulae. Through processes of bioinformationalisation, life, in all of its forms, becomes transmuted into anonymous digital data that is trackable and that can be killed extrajudiciously, that is to say, with impunity. I say 'anonymous' precisely because the US often kills targets whose identities are not known. Geolocation technology, the DoD says, has 'cued and compressed numerous "kill chains" (i.e., all of the steps taken to find, track, target, and engage the enemy)' (Scahill and Greenwald 2014). The compression of the drone kill chain has been enabled by the often critical conflation of a cell phone with the unknown identity of the user—in the words of the above-cited drone operator, 'We're not going after people—we're going after their phones, in the hopes that the person on the other end of that missile is the bad guy.'

The investigative journalist, Jeremy Scahill, provides in his book, *Dirty Wars* (2013), a powerful exposé of the covert wars that the US is conducting through its drone campaign. He explains what is at stake in this drone targeting program: 'In some cases, the specific individuals are being targeted, even though the United States doesn't know their identities, and may not have any actual evidence that they're involved in terrorist activity' (cited in Channel 4 2014). Operating under the dubious rubric of exercising its right to self-defence in response to an imminent threat, the 'US believes determining if a terrorist is an imminent threat "does not require the United States to have clear evidence that a specific attack on US persons and interest will take place in the immediate future"' (Serle 2014). The category of 'imminent threat' is inbuilt with an extraordinary latitude in terms of the subjects it enables the US military to target. It has resulted in the deaths of innumerable subjects whose names are unknown and who, when they are finally identified, are found to have no connections at all to such targeted groups as al-Qaeda. In its analysis of 400 US drone strikes in Pakistan alone, the Bureau of Investigative Journalism found that 'fewer than 4% of the people killed had been identified by available records as named members of al-Qaeda. This calls into question US Secretary of State John Kerry's claim last year that only "confirmed terrorist targets at the highest levels were fired at'" (Serle 2014). A recent report by Reprieve (2014) has brought to light that up to 874 'unknowns' have been killed by US drone strikes in the hunt for 24 targeted individuals. The Reprieve report documents the extraordinary toll on civilians that has been exacted by these metadata drone kills. It estimates that '96.5% of casualties from US drone strikes are civilians' (Dvorin 2014). Reprieve's Jennifer Gibson, who led the study, elaborates on the meaning of these statistics: 'Drone strikes have been sold to the American public on the claim that they're "precise." But they are only as precise as the intelligence that feeds them. There is nothing precise about the intelligence that results in the deaths of 28 unknown people, including women and children, for every "bad guy" the US goes after' (cited in Khan 2014). The Reprieve report documents the manner in which certain targeted individuals have been listed as having been killed up to six times, with the result that dozens of unknown civilians have actually been killed by the time the reporting process authenticates a targeted strike.

The geolocation technology's foundational dependence on an algorithmic formula provides a calculus of risk probability for a designated target whose identity remains unknown. In other words, this algorithmic program works to transmute difference into serial sameness and interchangeability. Knowledge, in this scientific schema, is what Friedrich Nietzsche would term as 'the falsifying of the multifarious and incalculable into the identical, similar, and calculable' (cited in Babich 1994, p. 102). What is operative here is the serial conflation of a technological signature with the unknown identity of the user of the cell phone. The knowledge of the one is rendered as interchangeable with the non-knowledge of the other. The contours of this convoluted scientific epistemology are clarified by Babette Babich in her critique of contemporary science. 'Today's science,' she writes (p. 199), 'plays on the limits of knowledge, it exploits what it knows and uses it as a template for what it does not know.' In other words, once situated in the context of a geolocation drone cue and kill program, the knowledge of a cell phone's electronic signature is transposed as a template onto the unknown identity of the phone's user in order to render the subject 'knowable.' 'The claim to absolute truth made by contemporary science,' Babich (p. 199) underlines, 'is prudently approximative, asymptomatic, a peripheral movement: one does not claim to have the truth, and this is one's claim to the truth.' In the practice of geolocation drone kills, the identity of the human drone target is often approximative with their cell phone; if it is revealed that the human target killed by a drone strike was based on a mistaken identity, this error is scripted as asymptomatic in the scheme of things and as peripheral to the larger concerns of winning the war on terrorists. 'In this way science need not acknowledge its dissembling tactics. What is known and what is not known are thus continuously connected' (p. 199). Nowhere, perhaps, is this truth more graphically evidenced than in the US geolocation drone kill program, where what is known is inextricably and often fatally connected to what is unknown.

This interplay between the known and unknown, between probability and chance, effectively constitutes what I elsewhere term 'drone casino mimesis' (Pugliese 2016). Drone casino mimesis is exemplified by such drone Ground Control Stations as Nellis and Creech Air Force Bases, both located just outside the gambling capital of Las Vegas. The conduct of drone kills from such sites brings into focus the crossover between video-gaming practices, gambling, and the speculative nature of tele-mediated drone kills. Drone casino mimesis identifies the agentic role of casino and gaming technologies precisely as 'actors' (Latour 2004, p. 226) in the shaping and mutating of both the technologies and conduct of war. Situated within the configuration of drone casino mimesis, I contend that the mounting toll of civilian deaths due to drone strikes is not only a result of human failure or error—for example, the misreading of drone video feed, the miscalculation of targets and so on. Rather, civilian drone kills must be seen as an in-built effect of digital kill technologies that are underpinned precisely by both the morphology (gaming consoles, video screens, and joysticks) and the algorithmic infrastructure of gaming—with its foundational dependence on 'good approximation' ratios and probability computation.

In proceeding to analyse the bioinformationalisation of life, I want to disclose the ways in which the US DoD's increasing reliance on death by metadata is inscribed by a biopolitics that is predicated on anthropocentric hierarchies of life. In the course of this chapter, I will situate the US' digitised surveillance and war technologies that kill by metadata in the geopolitical contexts of Pakistan and Yemen. The manner in which these surveillance and war technologies are foundationally informed by both the biopolitical caesura (with its violent division between human and animal) and a virulent anthropocentrism is powerfully captured by the words of the elderly mother of a US drone strike victim. The drone strike in question occurred on 12 December 2013, killing 12 people in a wedding procession in Yemen. In flagrant denial of the testimonies delivered by the survivors of this drone attack, 'The Pentagon's Joint Special Operations Command, which carried out the December strike, insists that everyone killed or wounded in the attack was an Al Qaeda militant and therefore a lawful military target' (Dilanian 2014). In the wake of the fatal strike, the elderly mother says: 'Whatever we do, they will never look at us as human beings ... We end up with wounds they cannot see' (cited in Drones Team 2013). Two things are worth noting here. Firstly, that Yemeni civilians are systemically precluded from occupying the category of human subjects when caught in the crosshairs of drone surveillance and killing technologies. As such, they are rendered as non-human animals that can be dispatched with the swivel of a joystick from drone Ground Control Stations situated thousands of miles away on the US mainland. Secondly, post the drone strike, the trauma and suffering unleashed by the firing of drone Hellfire missiles into civilian populations becomes entirely occluded and effaced for the very agents who have conducted these drone kills.

In articulating that 'Whatever we do, they will never look at us as human beings,' this Yemeni civilian draws our attention to the mediating forces that inform and calibrate those very scopic technologies trained on their Yemeni targets. Operative here are the disembodying effects of the bioinformationalisation of life. Through the deployment of a series of algorithms and the compounding biopolitical violence of a racio-speciesism, the target is configured as a lesser form of life in the scale of the anthropocentric hierarchy (see Pugliese 2013, pp. 41–2).

Geobiomorphology: The Transliteration of Algorithms to Flesh

In the course of an interview about a US cruise missile strike in al-Majalah, Yemen, Scahill describes the account of the survivors regarding the toll of this strike: 'I talked to tribal leaders who went there within 24 hours of the strike, and they described a scene where livestock and humans—the flesh of livestock and humans was melted together, and they couldn't determine if it was goats or sheep or human flesh, and they were trying to figure out how even to bury their dead' (cited in Democracy Now! 2013). These accounts of human and animal flesh inextricably bound in the moment of violent death caused by US drone or cruise missile strikes are to be found across the numerous investigative reports that document the civilian toll of these attacks. In a statement to the United States Senate Judiciary Committee, Subcommittee on the Constitution, Civil Rights, and Human Rights on the US drone wars, Farea Al-Muslimi, a Yemeni farmer, describes the strike by a US cruise missile on the village of al-Majalah, Yemen, on 17 December 2009:

> In the poor village that day, more than 40 civilians were killed, including four pregnant women. Bin Fareed was one of the first people to the scene. He and others tried to rescue civilians. He told me their bodies were so decimated that it was impossible to differentiate between the children, the women, and their animals. Some of these innocent people were buried in the same grave as animals. (Al-Muslimi 2013)

In these profoundly moving accounts, humans and animals are fused into a composite residue of inextricable flesh through the violence of war. The one melts into the other. The one is buried with the other. Taking these harrowing accounts of fused flesh as my point of departure, I want to proceed to delineate the contours of an ethical ground that would encompass both the human and the animal as grievable forms of life in the face of the violence of war. I can think of no more appropriate way of doing this than by *enfleshing* the world via the work of Maurice Merleau-Ponty. In conceptualising our relation to the world, Merleau-Ponty (1997, p. 127) posits the flesh as that which conjoins one to the other: 'The presence of the world is precisely the presence of its flesh to my flesh.' Merleau-Ponty's theorising of flesh disrupts the circumscriptions of the anthropocentric frame. Flesh emerges, for him, as both a general and specific modality of being in the world. It is a general modality as flesh signifies the condition of possibility of being in the world. It is specific as the particular subject is always-already enfleshed in its coming into being; the moment of death enunciates the inexorable falling away of the flesh and its gradual dissolution into something other. Before its disintegration into ashes or dust, the flesh of the dead still persists in signifying a binding, if transient, relation to the world. In the melting of human and animal flesh after the violence of a missile strike, there emerges a binding of one to the other that is predicated on the material fact that 'there is this thickness of flesh between us' (p. 127). The thickness of this flesh is what conjoins one victim to the other and what, post a missile strike, cannot be categorically divided and neatly assigned along the biopolitical hierarchy of life. Through this thickness of flesh, human and animal, subject and object, figure and ground dissolve. The thickness of this flesh is also what evidences the materiality of the remainders of the dead: in its decomposing and shredded corporeity, this thickness of flesh is what forensically testifies to the act of fatal violence. It exposes to view what is virtually everywhere denied by the US administration: the fact that its missile strikes are neither surgical nor accurate and that they kill much more than their designated "suspect" targets. If these strikes can be called "surgical" in any way, it is through their violent capacity to slice up living entities and to reduce them to scattered fragments of undifferentiated flesh.

In the wake of the violence unleashed by the missile blast, there is a denucleation of the identity of the one and the other. Following a drone strike in the Yemeni village of Al-Shihr, Hassan Ibrahim Suleiman describes how the victims' 'bodies were shredded. We collected the remains without knowing who they were' (cited in Alkarama Foundation 2013). The flesh of the one is so fused into the flesh of the other as to be reconstituted into an indeterminate and horizontal consanguinity. Articulated here, in this binding of different flesh, is 'the relation of the human and animality ... not [as] a hierarchical relation, but lateral, an overcoming that does not abolish kinship'—on the contrary, it works to materialise what Merleau-Ponty (2003, p. 268, 271) terms a 'strange kinship.' Crucially, for Merleau-Ponty (1997, p. 272), this kinship between the flesh of animals and humans is ethically envisaged as constituted at once by divergence (*écart*) or difference and a relational similarity: the qualifier 'strange' in this kinship relation works to underscore unassimilable difference or alterity in the context of binding relations of consanguinity. His concept of *intercorporeity*, as the relational constitution of bodies, is what 'founds transitivity from one body to another' (p. 141, 143).

In the harrowed ground of a lethal strike, the impossibility of unbinding this encrypted consanguinity calls for the burial of one with the other: 'it was impossible to differentiate between the children, the women, and their animals. Some of these innocent people were buried in the same grave as animals' (Al-Muslimi 2013). In the wake of these drone massacres of civilians and animals, one can discern a perverse relation of correspondence between the metadata program and its resultant "outputs." In the first instance, death by metadata involves the killing of subjects whose identities are not known. The killing violence that is unleashed by the drone's Hellfire missiles works, in turn, to render the very identity of the targets unknowable to the families and villagers who are left to collect their anonymous, shredded remains.

Sheikh Saleh bin Fareed, in his testimony to Human Rights Watch, describes the post-drone strike scene he was compelled to witness:

> Goats, sheep, cows, dogs, and people, you could see their bodies scattered everywhere, some many meters away. The clothes of the women and children were hanging from the treetops with the flesh on every tree, every rock. But you did not know if the flesh was of human beings or animals. Some bodies were intact but most, they melted. (cited in Human Rights Watch 2013, p. 77)

Merleau-Ponty's conceptualisation of the 'flesh of the world' here assumes an entirely other dimension of meaning. In the context of this site of saturated violence, his concept becomes critically resignified. Reading this scene of carnage, I become a tertiary witness to 'a flesh of things' (Merleau-Ponty 1997, p. 248, 133). A flesh of things articulates the shredding of human and animal flesh by a missile and its transmutation into a thing that literally *enfleshes* the world: hanging from treetops and rocks, human-animal flesh bears witness, as dispersed fragments, to the rendering of lives into the mere 'tissue of things' (p. 135). In such moments there are not indivisible beings, but intercorporeal assemblages of flesh where individual beings have been rendered into undifferentiated biological substance. Animals and humans are, in this site of explosive decomposition, situated beyond the taxonomic and hierarchical domains of species differentiation. Collectively, they embody not different species but 'families of trajectories' (Merleau-Ponty 2003, p. 93) that, after the violence of the missile's explosion, meld into each other, even as they are compelled to assume the contours of the landscape. Citing F. London and E. Bauer's work on quantum physics, Merleau-Ponty (2003, p. 92) describes how a 'theory of species' can in effect dissolve into 'an indiscernibility of particles of the same species.' One can, indeed, name this scene of violence as a site that visibly instantiates the indiscernibility of particles of different species through a process of fissile dissolution and recombination.

The violent dissolution and recombination of life is, in the context of such sites of saturated violence, articulated through the register of melted flesh that becomes entirely coextensive with its *geocorpography*. In coining the term geocorpography some years ago, I wanted to bring into focus the impossibility of disarticulating the body from its geopolitical locus, and to materialise the multiple significations that accrue from this understanding of the geo-corporeal nexus (Pugliese 2007). In the context of the massacre that was perpetrated by the US in this Yemeni village, geocorpography enunciates the violent enmeshment of the flesh and blood of the body with the geopolitics of war and empire. This site of carnage, however, takes the concept of geocorpography to another level of signification. It compels the coining of a new composite term that acknowledges the specific modality of this violence: *geobiomorphology*. The bodies of humans and animals are here compelled to enflesh the world through the violence of war in a brutally literal manner: the dismembered and melted flesh becomes the 'tissue of things' as it geobiomorphologically enfolds the contours of trees and rocks. What we witness in this scene of carnage is the transliteration of metadata algorithms to flesh. The abstracting and decorporealising operations of metadata 'without content' are, in these contexts of militarised slaughter of humans and animals, geobiomorphologically realised and grounded in the trammelled lands of the Global South.

I want to pause for a moment to flesh out the significance of my choice of the term *transliteration*. Oriented by Susan Stryker's (2006, pp. 1–18) groundbreaking work on the critically interrogative power of *trans*, I want to examine this term's range of significations and systems of relations once it is transposed to the domain of the military-industrial-surveillance complex and its attendant necropolitical arsenal and effects. Precisely as a term that is animated by a transitive prefix, this term establishes critical points of processual connection between metadata and flesh. Transliteration figures forth the spatio-temporal process encompassed by the necropolitical assemblage of drone kills. The term draws attention, through its transitive semantics, to the shift from the spaces of a drone Ground Control Station in some US metropolitan city, such as Las Vegas, to the killing fields of the Global South, as the fabled "ungoverned" lands that abut onto the extra-territorial edges of US empire and that are invariably positioned as posing "imminent threats" to its national interests. The term also materialises the tele-techno mediated temporal shifts that establish fundamental contra-distinctions between the insular time of execution, from the safety and comfort of an ergonomically-designed drone pilot's cubicle, to the time of utter exposure and fatal violence experienced in the killing field of a missile blast. Transliteration marks the process of converting an abstracted medium of numerals, algorithmic formulae and pixels to the earthly medium of flesh. It names the conversion of metadata's mathematico-scientific formulae to the flesh of the world. Let me stress that what I am not positing here is the concept of flesh as some purely natural biological substance. On the contrary, I understand flesh to be always-already mediated by the very somatechnic processes that render it culturally intelligible precisely *as flesh*. What I am drawing attention to is a different order of mediation fundamentally predicated on a militarised series of instrumentalising functions designed to render the flesh of the Global South into little more than disposable biological substance ontologically abstracted from its target human-animal subjects. In other words, what we have here is the following seeming paradox: the drone targets of the Global South are, as mere 'patterns of life,' 'no-bodies,' to draw on Denise Ferreira da Silva's (2009, p. 220) appurtenant term. Networked through the abstracting schema of metadata, these no-bodies are configured as nothing less than generic, anomic, and wholly killable flesh.

In elaborating on the 'flesh of the visible,' Merleau-Ponty (1997, p. 136) writes that, by that concept, 'we mean that carnal being, as a being of depths, of several leaves or several faces.' In the wake of the drone strike in this Yemeni village, the flesh of the visible becomes something else altogether: the human and animal beings who are its victims become, through the force of explosive shredding, deprived of the depth of their being, precisely as they are transmuted into surface fragments, leaves of flesh without faces that sway from the treetops. This, then, is the geobiomorphology of a drone kill, as the abstract formulae that deliver to drone pilots 'suspect patterns of life' are transliterated into flesh: the wounds in the earth are the mouths of the dead; their flesh is strange fruit that hangs from leaves and branches; their hair, hide, and shards of bone trace the lineaments of the debris field; and their blood pools into scarlet soaks.

Reading Merleau-Ponty in the necropolitical context of this massacre, I am compelled to transpose his celebratory writings on flesh of the world to the field of war and, consequently, his writings proceed to signify a range of unintended meanings and a/effects. In theorising the flesh of the world, he writes (1997, p. 114): 'The space, the time of things are shreds of himself, of a multiplicity of individuals synchronically and diachronically distributed, but a relief of the simultaneous and of the successive, a spatial and temporal pulp.' The animals and humans that are the victims of this US drone kill are, post the missile strike, transmuted into the dead time of things in which the shreds of their flesh constitute a multiplicity of individuals that is synchronically and diachronically distributed. This is flesh that is synchronised in the killing instance of the strike: in the wake of the force of the blast, a diachronic distribution of the shredded flesh unfolds. Caught in the violent simultaneity of the fatal moment, the victims' flesh is then diachronically inscribed through the successive acts of dispersal and funereal gathering. The anomic genericity of suspect and targeted 'patterns of life' is, in the instant of the killing strike, transliterated into de-individuated fragments of flesh without bodies, names, or identities. I can think of no more accurate way of naming what has transpired in this site of US military carnage than the rendering of its human and animal victims into 'a spatial and temporal pulp.' One can no longer talk of corporeality here. Post the blast of a drone Hellfire missile, the corpora of animals-humans are rendered into shredded carnality. In other words, operative here is the dehiscence of the body through the violence of an explosive centripetality that disseminates flesh. The moment of lethal violence transmutes flesh into unidentifiable biological substance that is violently compelled geobiomorphologically to assume the topographical contours of the debris field.

Reflecting on the manner in which flesh must be seen 'as "element," in the sense it was used to speak of water, air, earth, and fire, that is, in the sense of a *general thing*,' Merleau-Ponty (1997, pp. 139–40) concludes that what this conceptualisation establishes is 'the inauguration of the *where* and the *when*, the possibility of exigency for the fact; in a word: facticity—what makes the fact be a fact. And at the same time, what makes the facts have meaning—makes the fragmentary facts dispose themselves about "something."' Situated in context of this Yemen killing field, the flesh of the human and animal victims becomes an element in the forensic ecology of the place: it is inextricably bound to the earth, air, and trees of the site. As trace evidence of this US massacre, the fragmented flesh bears witness to the *where* and the *when* of the killing. As trace evidence dispersed across the scene of this massacre, it signifies the very facticity of the crime. These shards of lives, winnowed by the force of a missile blast guided by an abstracting algorithm designating a suspect 'pattern of life,' emerge as the forensic traces of a criminal act that will yet, because of the violent asymmetry of North/South power, evade any criminal prosecution.

Sheikh Saleh bin Fareed offers another perspective on the drone strike massacre in al-Majalah:

'When we went there, we could not believe our eyes. I mean, if somebody had a weak heart, I think he would collapse. You see goats and sheep all over, you see the heads of those who were killed here and there. You see their bodies, you see children.' … Body parts were strewn around the village. 'You could not tell if this meat belongs to animals or human beings,' he remembered. They tried to gather what body parts they could to bury the dead … As bin Fareed surveyed the carnage, most of the victims he saw were women and children. 'They were all children, old women, all kinds of sheep and goats and cows. Unbelievable.' (cited in Scahill 2013, p. 305)

In surveying the site of this massacre, bin Fareed talks not of flesh but of 'meat' as that which remains after the missile blast. The use of the term

'meat' brings into focus the fact that the consanguinity that I drew attention to above is emblematised by the manner in which the militarised
violence of the Global North works to transmute the living subjects of
the Global South into what I have elsewhere termed as *carcasses* (Pugliese
2013, p. 166–9). The biopolitical caesura, through its human/animal
division, renders the civilians killed by the West in the course of the war
on terror, in all of its manifold incarnations, as so many animal carcasses
that, in effect, do not die but merely perish. Inscribing this Western production of human carcasses from the Global South is the metaphysics
of a virulent (racio-) anthropocentrism that finds its clinical articulation
in Heidegger. For Heidegger (1995, p. 267), 'the animal,' because it is
defined by a fundamental series of privations and captivations, 'cannot
die in the sense in which dying is ascribed to human beings but can only
come to an end.' 'To die,' Heidegger (1975, p. 178) elaborates, 'means to
be capable of death as death. Only man dies. The animal perishes.' The
violent operations of racio-speciesism render the subjects of the Global
South as non-human animals captivated in their lawlessness and inhu-
man savagery and deficient in everything that defines the human-rights-
bearing subject. In contradistinction to the individuating singularity
of the Western subject as named person, they embody the anonymous
genericity of the animal and the seriality of the undifferentiated and
fungible carcass. As subjects incapable of embodying the figure of "the
human," they are animals who, when killed by drone attacks, do not die
but only come to an end. In Western-mediated contexts, what remains is
the carcass that is not worthy of mourning and that, as carcass that merely
perishes, need not be taken into account as a human death. Mohamed al-
Qawli, a Yemeni civilian, articulates as much when he attempts to make
sense of the carnage he was compelled to witness following a US drone
strike: 'All of us in the village heard a large explosion … We picked up
the burned body parts. They were all over [the place] … My brother was
completely charred. We identified him by his teeth. It's as if they killed
animals' (cited in Bayoumy 2013). Situated in this racio-anthropocentric
context, the brazen denial by the US administration of the thousands
of drone fatalities it has caused through its death-by-metadata program
assumes its own internal logic: No one dies in these US drone strikes.
Only animals perish.

Note

1. In his work, Michel Foucault insistently draws attention to the manner in which relations of power are inextricably tied to the production of knowledge. Contesting merely negative understandings of power, Foucault brings into focus its productive dimensions. Power, Foucault (1980, p. 119) writes, 'needs to be considered as a productive network that runs through the whole social body.' Across the corpus of his work, he evidences the manner in which power works to produce bodies of knowledge invested with truth and authority. In the context of this chapter, military relations of power are exercised through a range of scientific knowledges that work to instrumentalise, through abstracting software and remote killing technologies, the human and animal subjects targeted by US drones.

References

Alkarama Foundation. (2013). *The United States' War on Yemen: Drone attacks. Report submitted to the special Rapporteur on the promotion and protection of human rights and fundamental freedoms while countering terror*, 3 June. http://en.alkarama.org/documents/ALK_USA-Yemen_Drones_SRCTwHR_4June2013_Final_EN.pdf. Accessed 21 June 2013.

Al-Muslimi, F. (2013). *United States Senate Judiciary Committee, Subcommittee on The Constitution, Civil Rights and Human Rights, Drone Wars: The constitutional and counterterrorism implications of targeted killing. Statement of Farea Al-Muslimi*, 23 April. http://www.emptywheel.net/wp-content/uploads/2013/04/04-23-13Al-MuslimiTestimony.pdf. Accessed 1 July 2013.

Babich, B. E. (1994). *Nietzsche's philosophy of science*. Albany: State University of New York Press.

Bayoumy, Y. (2013). Al Qaeda gains sympathy in Yemen as US drone strikes wedding party. *Business Insider*, 10 December. http://www.businessinsider.com.au/al-qaeda-gains-sympathy-in-yemen-as-us-drone-strikes-a-wedding-party-2013-12. Accessed 11 Dec 2013.

Channel 4 News. (2014). What is ABI – And will it make drones even deadlier? 22 February. http://www.channel4.com/news/drones-abi-activity-based-intelligence-yemen-strike-attack. Accessed 24 Feb 2014.

Cole, D. (2014). We kill people based on metadata. *New York Review of Books*, 10 May. http://www.nybooks.com/blogs/nyrblog/2014/may/10/we-kill-people-based-metadata/. Accessed 22 May 2014.

da Silva, D. F. (2009). No-bodies: Law, raciality and violence. *Griffith Law Review, 18*(2), 212–236.

Democracy Now! (2013). Yemeni reporter who exposed U.S. drone strike freed from prison after jailing at Obama's request. *Democracy Now!*, 25 July. http://www.democracynow.org/2013/7/25/yemeni_reporter_who_exposed_us_drone. Accessed 26 July 2013.

Dilanian, K. (2014). Debate grows over proposal for CIA to turn over drones to Pentagon. *LA Times*, 11 May. http://touch.latimes.com/#section/-1/article/p2p-80167118/. Accessed 17 May 2014.

Drones Team. (2013). Get the data: Drone wars: Yemen: Reported US covert actions. *The Bureau of Investigative Journalism*, 3 January. http://www.the-bureauinvestigates.com/2013/01/03/yemen-reported-us-covert-actions-2013/. Accessed 6 Jan 2013.

Dvorin, T. (2014). 96.5% of casualties from US drone strikes civilians. *Arutz Sheva/Israel International News*, 26 November. http://www.israelnational-news.com/News/News.aspx/187922. Accessed 30 Nov 2014.

Foucault, M. (1980). *Power/knowledge: Selected interviews and other writings 1972–1977*. New York: Pantheon Books.

Heidegger, M. (1975). *Poetry, language, thought*. New York: Harper and Row.

Heidegger, M. (1978). The question concerning technology. In D. F. Krell (Ed.), *Martin Heidegger: Basic writings*. London: Routledge and Kegan Paul.

Heidegger, M. (1995). *The fundamental concepts of metaphysics: World, finitude, solitude*. Bloomington: Indian University Press.

Human Rights Watch. (2013). *'Between a drone and al-Qaeda': The civilian cost of US targeted killings in Yemen*. http://www.hrw.org/sites/default/files/reports/yemen1013_ForUpload.pdf. Accessed 12 Nov 2013.

Khan, H. (2014). "Precise" drone strikes in Pakistan: 874 "Unknowns" killed in US hunt for terrorists. *Express Tribune*, 26 November. http://tribune.com.pk/story/797295/precise-drone-strikes-in-pakistan-874-unknowns-killed-in-us-hunt-for-24-terrorists/. Accessed 27 Nov 2014.

Latour, B. (2004). Nonhumans. In S. Harrison, S. Pile, & N. Thrift (Eds.), *Patterned ground: Entanglements of nature and culture*. London: Reaktion Books.

Merleau-Ponty, M. (1997). *The visible and the invisible*. Evanston: Northwestern University Press.

Merleau-Ponty, M. (2003). *Nature: Course notes from the Collège de France*. Evanston: Northwestern University Press.

Pugliese, J. (2007). Geocorpographies of torture. *Critical Race and Whiteness Studies, 3*(1). http://www.acrawsa.org.au/. Accessed 23 Sept 2014.

Pugliese, J. (2013). *State violence and execution of law: Biopolitical caesurae of torture, black sites, drones.* Abingdon/New York: Routledge.

Pugliese, J. (2016). Drone casino mimesis: Telewarfare and the militarisation of civilian sites, practices and technologies. *Journal of Sociology, 53*(3), 500–521.

Reprieve. (2014). You never die twice: Multiple kills in the US drone program. *Reprieve,* 24 November. http://www.reprieve.org/uploads/2/6/3/3/26338131/2014_11_24_pub_you_never_die_twice_-_multiple_kills_in_the_us_drone_program.pdf. Accessed 25 Nov 2014.

RT. (2014). Former CIA director: "We kill people based on metadata." *RT,* 12 May. http://rt.com/usa/158460-cia-director-metadata-kill-people/. Accessed 14 May 2014.

Scahill, J. (2013). *Dirty wars: The world is a battlefield.* New York: Nation Books.

Scahill, J., & Greenwald, G. (2014). The NSA's secret role in the US assassination program. *The Intercept,* 10 February. http://firstlook.org/theintercept/2014/02/10/the-nsa-secret-role/. Accessed 12 Feb 2014.

Serle, J. (2014). Only 4% of drone victims in Pakistan named as al Qaeda members. *Bureau of Investigative Journalism,* 16 October. http://www.thebureauinvestigates.com/namingthedead/only-4-of-drone-victims-in-pakistan-named-as-al-qaeda-members/?lang=en. Accessed 17 Oct 2014.

Stryker, S. (2006). (De)subjugated knowledges: An introduction to transgender studies. In S. Stryker & S. Whittle (Eds.), *The transgender studies reader.* New York: Routledge.

2

Of Bodies, Borders, and Barebacking: The Geocorpographies of HIV

Joshua Pocius

Introduction

Sex and pleasure are always constituted in relation to boundaries. From the axiomatic notion that (penetrative) sex inherently involves the transgression of physical borders between the Self and the Other, to the technologies of containment such as condoms that seek to limit the potential for corporeal coalescence, to the designation of spatial zones in which certain kinds of embodied sexual practice may or not be permissible, borders and boundaries play a significant role in erotic life. In the context of the ongoing human immunodeficiency virus/acquired immune deficiency syndrome (HIV/AIDS) pandemic, the cultural mediation of AIDS histories and gay male pornographic representation are juxtaposed against competing discourses around the use of biomedical technologies in the service of public health and/or erotic pleasure. Tantamount to

J. Pocius (✉)
School of Culture and Communication,
University of Melbourne, Melbourne, VIC, Australia

© The Author(s) 2016
H. Randell-Moon, R. Tippet (eds.), *Security, Race, Biopower*,
DOI 10.1057/978-1-137-55408-6_2

these discourses are the ways in which embodied subjects and erotic practices are circumscribed by borders which determine the proximity to biomedical interventions into pleasure and protection. In this chapter, I situate the emergence of pre-exposure prophylaxis (PrEP) as a biomedical technology which straddles discourses of public health and erotic pleasure. Addressing the first fictionalised screen-mediated representation of PrEP—a scene from a gay male pornographic film—I locate PrEP's emergence within the context of HIV/AIDS cultural production in 2014 as an indication of a temporal and generational shift within gay male sexual cultures. Characterising PrEP as an example of 'informed matter' (Rosengarten 2009), I argue that the mobilisation of biomedical technologies launched under an official public health discourse and subsequently mediated through gay male sexual cultures as an erotic enabler illustrates what I term the eRotics of HIV. Finally, I assert that the emergence of PrEP as a sanctioned method of HIV prevention in certain geopolitical contexts is concomitant with a broader regime of demarcation in which bodies and their proximity to HIV and antiretroviral therapy are geopolitically constituted. What transpires from the confluence of an eRotics of HIV amidst the context of cartographic violence constitutes a geocorpography of HIV in which certain geopolitically-constituted subjects are granted immunity from the potential risks harboured by pleasures of the flesh through the same logic which limits antiretroviral access for those who already require it.

HIV/AIDS Cultural Production in 2014

The year 2014 was marked by a number of notable developments concerning HIV/AIDS, both cultural and epidemiological. Three major films addressing the North American AIDS crisis in the 1980s were in cinemas and on television. Chris Mason Johnson's film *Test* reimagined the San Francisco contemporary dance scene at the emergence of the first HIV-antibody test, invoking an affective nostalgia both for a San Francisco long-since lost to "post-AIDS" tech-boom gentrification and for the purgatorial self-reflexivity of a two-week wait for test results. Jean-Marc Vallee's Hollywood blockbuster film *Dallas Buyers Club* portrayed the clandestine networks of people living with AIDS (PWAs) illegally importing and distributing experimental HIV drugs into the United States. Ryan Murphy's Home Box Office (HBO) adaptation of Larry Kramer's 1984 play *The Normal Heart* depicted early AIDS activism in the gay community in New York City and Kramer's advocation for a shift in gay male sexual culture in order to stem the epidemic. Viewed together, these films constitute the emergence of a specific form of AIDS nostalgia that reimagines the spaces and cultural temporalities of the AIDS epidemic era. Yet this 'backwards gazing' (Love 2007) and memorialisation of the North American AIDS crisis also obfuscates the contemporary lived realities of the ongoing HIV/AIDS pandemic, furthering the narrative that AIDS was something that 'happened then' or something that 'happens over there'.

2014 was also a noteworthy year for developments in HIV prevention technologies. In May of that year, the US Centers for Disease Control (CDC) altered the wording in the clinical guidelines for Pre-Exposure Prophylaxis (PrEP). The latest in a multi-pronged pharmacological intervention into the HIV pandemic, PrEP joins more familiar strategies such as Treatment-as-Prevention (TasP) and Post-Exposure Prophylaxis (PEP), which involve using Highly-Active Anti-Retroviral Therapy (HAART) in order to reduce the viral load in seropositives to an 'undetectable' level and prevent seroconversion following a suspected exposure respectively. PrEP also involves the use of HAART, however in this case it is used as a preventative measure for seronegative people, taken once per day. Whereas the CDC had approved the use of Gilead's Truvada combination antiretroviral for PrEP in 2012, in 2014 the CDC amended their approach from approving to *recommending* its use by those with certain HIV risk profiles (CDC 2014a).

Finally, 2014 was the year in which the notorious gay pornography studio Treasure Island Media released two films which pushed at the boundaries of HIV representation. In *Viral Loads* (dir. Paul Morris, P. 2014), performer Blue Bailey is inseminated with the contents of a jar labelled 'POZ CUM', denoting the alleged seropositivity of the semen in the container. Although Treasure Island Media have often hinted at the exploitation of the "bug chaser" narrative of eroticising HIV transmission in previous titles, this was the first instance of an explicit reference to the presence of the virus within the space of the diegesis. The second Treasure Island Media film released in 2014 garnered controversy for its cultural mediation of a new generation of gay men who apparently disavow the threat of HIV transmission. In the scene '#TeenageTruvadaWhore', which was pre-released from the film *London Uncut* (dir. Liam Cole 2014), 18-year-old Josh Taylor is shown being penetrated without condoms by seven older men. The scene is notable for being the first fictional screen-mediated representation of PrEP, and for the complications of cultural temporality that PrEP's adoption has induced. The scene is temporally framed as a break with the HIV eroticisation narratives exemplified in *Viral Loads*, self-consciously representing a post-HIV gay male sexual culture enabled by chemoprophylactic strategies of HIV prevention. In this sense, the scene illustrates what Paul Morris, founder of Treasure Island Media, suggests is the contemporary reality in which simple strategies like PrEP are able to 'render HIV a non-issue' (cited in McCasker 2014). Within the context of the idealised target consumer for PrEP—the urban, middle-class, male, gay 'neoliberal sexual actor' (Adam 2005) of the Global North—chemoprophylaxis does indeed raise the prospect of a condomless sexual future engendered through continual antiretroviral consumption, and in turn, a future in which fear of HIV transmission is coded as historical. Temporally distancing HIV from the present evokes a certain nostalgia for pre-AIDS gay sexual cultures in the Western gay cultural centres of New York City, San Francisco, Los Angeles, and Sydney. However, as I demonstrate in this chapter, access to antiretroviral therapies that prevent HIV from progressing into AIDS, let alone access to the sexual pleasure imagined by Morris and "unmediated" by latex, is overwhelmingly determined by the geocorpographies of HIV.

Emergent Chemoprophylaxis

As a new biomedical intervention aimed at reducing HIV transmission through an immunitarian paradigm which anticipates a future risk to exposure, PrEP marks a divergence from the conventional applications of antiretrovirals. Whereas post-exposure prophylaxis intervenes *after* an instance of risk, and whereas treatment-as-prevention intervenes in the body of the seropositive in order to reduce the risk they pose to seronegatives, the moral economy of PrEP situates the PrEP user in relation to embodied risky practices and operates on the assumption that the PrEP user will engage in practices of risk with or without access to PrEP. It is in this context that the CDC's alteration of its wording—from endorsing PrEP to *recommending* it—signals a dramatic shift in the biopolitics of HIV. Although the CDC were careful not to publish the actual number of individuals they were recommending to be more or less immunised through chemoprophylaxis, according to the risk schemas in their guidelines, the number is estimated to be between 300,000 and 500,000 HIV-negative people (CDC 2014a). PrEP trials have also commenced in Europe and in the Australian states of Victoria and New South Wales for men who have sex with men (MSM), producing significant debates in MSM and HIV prevention circles about individual, community, and government responsibility for preventing HIV transmission through embodied practices. Whilst the Australian government is yet to release specific guidelines around the use of PrEP, the AIDS Council of New South Wales (ACON) released guidelines in February 2015 detailing how HIV-negative people might go about accessing PrEP prior to official endorsement, by obtaining an 'off-label' prescription from a doctor and purchasing and importing generic versions of the drug Truvada from Indian or Canadian online pharmacies (ACON 2015).

Controversy surrounds the use of PrEP as a strategy to prevent HIV transmission from three fronts. Firstly, a few prominent individuals and organisations who were involved in activism in the AIDS crisis years, including activist and playwright Larry Kramer and the head of the Los Angeles-based AIDS Healthcare Foundation (AHF) Michael Weinstein, have been outspoken in warning against adoption of a new pharmacological strategy for HIV prevention. Weinstein's well-funded AHF have launched a vitriolic and at times dishonest campaign against PrEP, taking out full-page newspaper advertisements, billboards, and accessing considerable mainstream media airtime as the voice of gay dissent in the PrEP debate. One such campaign, running in 11 print publications across California and Florida and titled 'PrEP Facts', misrepresented the efficacy results from major trials to suggest that PrEP was less effective than reported by the federal government by ignoring the impact of rates of adherence to efficacy results (Heywood 2014). Speaking to the Associated Press, Weinstein referred to Truvada as a 'party drug' (Crary 2014). Kramer has similarly been outspoken against the use of PrEP. Speaking to *The New York Times* in May of 2015, during the media hype preceding the HBO debut of Ryan Murphy's film-adaptation of Kramer's play, *The Normal Heart*, he said:

> Anybody who voluntarily takes an antiviral every day has got to have rocks in their heads …There's something to me cowardly about taking Truvada instead of using a condom. You're taking a drug that is poison to you, and it has lessened your energy to fight, to get involved, to do anything. (cited in Healy 2014)

The second controversy surrounding PrEP stems from the discourse and counter-discourse of the "Truvada Whore", a derogatory term originating amongst US gay male communities to stigmatise early adopters of PrEP that refers to the brand name of the only drug currently approved for use in pre-exposure prophylaxis. The form of this stigmatisation parallels the stigmatisation that many HIV-positive people face, inscribing their sexual embodiment of certain practices with value-laden judgements that afford levels of deservedness to their sexual health status. In the case of those who opt to take Truvada as a form of protection *against* acquiring HIV, the identity position of "Truvada user" has often been portrayed in terms of breaking with norms of gay sexual hygiene that emerged as safer-sex strategies that were relatively successful in stemming the spread of HIV during the years of the crisis but appear to be falling out of favour: this is illustrated by Kramer's assertion that PrEP users are 'cowardly' by opting for a chemoprophylactic route rather than using physical barriers such as condoms.

Thirdly, chemoprophylaxis has garnered significant controversy due to the nature of its deployment in mostly white, gay, male, urban, middle, and upper-class contexts in wilful ignorance of the broader biopolitical and necropolitical paradigms at play in the global frame of antiretroviral access. Truvada is a patented drug from the US pharmaceutical company Gilead Sciences, and is priced at US$1300 per month. Whilst the majority of these costs are currently funded by health insurance providers, state-funded Medicaid services, or by the company itself, the cost is an obvious barrier for a large proportion of global citizens who could benefit from the drug. Further, in the context of the neoliberal sexual actor and the increasingly neoliberal landscape of health provision in which the individual is expected to take greater responsibility for the provision of their healthcare regardless of structural and societal inequalities within the medico-juridical economy, the strategy of PrEP adoption as a form of HIV prevention points to concerns regarding the marketisation of healthcare and the pharmaceutical bondage of so-called "risk groups". Should PrEP become a widespread strategy for prophylaxis against HIV, akin to the contraceptive pill, this would represent an expansion of the potential consumer market for manufacturer Gilead Sciences to the tune of half a billion dollars per year (McNeil 2014). Yet this stands in stark contrast to the contemporary realities of access and uptake of HIV treatment under the inherently dysfunctional and inequitable health system in the US, where it is estimated that less than 17 per cent of the 1.2 million seropositive people have private health insurance while 30 per cent do not have any health insurance (CDC 2014b), and only 40 per cent are engaged in HIV treatment (CDC 2011, p. 1621).

Whereas the official public health discourse concerning PrEP constructs this biomedical intervention as an additional element of a broader regime of risk-minimisation strategies including the use of condoms, and constructs the ideal consumer of PrEP within notions of individual responsibility, the first screen-mediated fictionalised representation of PrEP is illustrative of PrEP's capacity to be consumed as a 'party drug' or, in the very least, a biomedical intervention to enhance erotic pleasure.

#TeenageTruvadaWhore

The bareback pornography studio Treasure Island Media and its founder, Paul Morris, consistently emphasise the documentary and ethnographic nature of their works. Morris articulates his identification as an ethnographer in his manifesto *No Limits: Necessary Danger in Male Porn*, presented at the World Pornography Conference in Los Angeles in 1998, in which he states that 'the representation not only of the truth but also of the complexity of the truth … is a responsibility of porn, the sexually indexical documentary genre' (Morris 1998, n.p.). In *Unlimited Intimacy*, Tim Dean identifies the central problematic of the bareback porn genre in terms of authenticity. For Dean, it is of little surprise that the self-representation of the subculture would be in pornographic terms, because 'bareback sex seems to call for witnesses and thus to generate documentary evidence, as well as communal bonds' (2009, p. 104). Dean details numerous filmic, narrative, and aesthetic strategies employed by Morris and Treasure Island Media which constitute this reach to documentary realism, including the use of 'non-professional' actors, whom Dean stipulates are not actors, 'just participants', adding that the camera operators are often visible within the frame, either marked as cameramen by their comparative lack of nudity or at times actively participating, which is 'evidence not of technical incompetence but of this pornography's blurring the distinction between participants and witnesses, just as bareback sex tries to blur the boundaries that separate persons from each other' (p. 105). These filmic and aesthetic strategies are all present in '#TeenageTruvadaWhore'; however, they are joined by an additional narrative and stylistic element in the opening of the scene, with a very quick shot of an email open on a computer screen. In the logic of the narrative, we are to understand that the email was authored by the "bottom" in the scene. Although it is unlikely that the performer actually wrote the email, the culture of the Treasure Island Media fandom evidences the likelihood that the email itself is "authentic". It reads:

Hey Liam

I am a fan of your hot work with treasure island media. I have been jerking off to your videos since I was 12 and you are still the best. I hope you enjoy my pictures and maybe jerk a few loads over them, as I have over your videos.

If you are reading this, please I have a request for future videos. Please one day will you make a bareback gang bang scene with a bottom guy who is my age (I am 18). We are not all innocent twinks lol. I have been fooling around for a long time and now I am on Truvada and I am definitely a total #TruvadaWhore haha. I love to be fucked bareback and take cum in my ass, and I have never used condoms since I started getting fucked in this day and age. I know older guys and parents think condoms for gay sex because they have experienced the bad times and the years of condoms, but that is in the past and now there is a new generation of gay guys like me on Truvada and we don't even need the word bareback because we don't have to fuck any other way. Sex with no condom is just sex, like for straights my age because their girlfriends use birth control.

And believe me, although I am only 18 it is easy for me to find older guys to fuck and cum in my ass. So please make a video for young guys like me. It would be so hot to see a young guy my age being the bottom in a gang bang of sexy older men (from 25 to 40 years is my favourite age, but any big cock is good). You will be a big hit with our generation of porn fans!

Hugs, kisses and licks.

#TeenageTruvadaWhore

In addition to the stylisation of the title of the scene, employing a hashtag in reference to a social media campaign to destigmatise Truvada users, this email, as a narrative framing device, is striking for the way it temporally situates the author in relation to both the bareback "subculture" and the broader cultural epistemology of condomless gay anal sex. The gradual and consistent decrease in self-reported condom use—particularly among younger gay men—in recent years has been attributed to a 'deficit of apprehension' (Race 2009, p. 111) whereby gay men who came of age *after* the transition to the 'post-AIDS' era did not have the same shared cultural memory of the ravages of the AIDS epidemic as the previous generation (Shernoff 2006, p. 107). As such, the balance between risk and pleasure is mediated by pharmacopower. By not bearing witness to or having cultural memory of peers dying from AIDS, some younger men instead 'feel resentment and deprivation at the constraints of safer sex' (Blechner 2002, p. 29).

Regardless of its authenticity, the presence of the email encapsulates a particular emergent gay male sexual subject at a point of cultural flux. With the author framed as a young adult who '[has] never used condoms since [he] started getting fucked in this day and age', what is represented here is not merely the notion of 'condom fatigue' that has been used to explain the rise in condomless sex in older gay men (Adam et al. 2005), but rather, a younger generation of gay men who have commenced their sexual lives in an era of chemoprophylaxis. At the confluence of biomedical intervention, embodied practice, and generational identification, the author's intimated self-articulation illustrates what Kane Race describes as the 'reflexive mediation' between the official discourses of public health which are communicated to gay men and the 'embodied habits' which eventuate (2003, p. 377). By situating himself and his generation in temporal contrast to 'older guys and parents' who have 'experienced the bad times and the years of condoms', the author is suggesting that condoms, as an element of a 'safe sex' strategy, are an archaic intrusion to sexual pleasure and practice and are no longer necessary in the context of chemoprophylaxis. The author disidentifies with the very word 'bareback' by rhetorically reframing the practice, suggesting that 'sex with no condom is just sex' and likening his use of PrEP to the way that straight people of his age use the contraceptive pill. In doing so, he is temporally situating his manifestation of condomless gay anal sex in a way that presents a striking contrast to the 'breeding culture' of 'serofraternity' discussed by Tim Dean (2008) and the notion that bareback practices constitute a discrete subcultural affinity. However, although the author disassociates himself from the loaded parlance of the term 'bareback' and its proximity to HIV and seroconversion, he still makes mention that he desires to be 'fucked' by a group of 'older men', insinuating an intergenerational connectedness through a shared sexual practice. In many ways, the author's temporal positionality points towards a sense of pre-AIDS nostalgia,[1] a notion that Morris supports, framing his work in a discussion with media studies scholar Susanna Paasonen as the production of a 'sense of continuity of a real and ages-old lineage … a living archive of male sexual practice' (cited in Paasonen and Morris 2014, p. 216). For Morris, the pharmaceutical strategies employed in managing the virus have created the very real possibility of rendering HIV a 'non-issue' (McCasker 2014). By making a concerted effort to afford HIV a level of visibility within a pornographic genre that is both entirely defined by the virus yet never makes mention of it, Morris is attempting to court a critical analysis on the part of the viewer and the broader public on the enormous shift in the meaning of HIV status since the years of the AIDS crisis. This perspective demonstrates the productivity of the interface of pharmacology and identitarian praxis, which I term an emergent eRotics of HIV.

My neologism of 'eRotics' plays on the typographic misreading of the symbol 'R̵', which is in widespread contemporary use in the medical profession as a short-hand for 'prescription' and as a symbol representing the pharmaceutical industry (Bailey 2008, p. 506). The origins of the symbol are a topic of debate (see Aronson 1999, p. 1543), however its misreading as 'R' (short for the Latin *recipe*, 'to take') in combination with 'x' offers the fruitful examination of the confluence of erotic embodied practices and the exoticisation of these practices, as enabled by neoliberal pharmaceutical developments. In this context, the pharmacological intervention of PrEP can be understood as an erotic enabler that offers "protected" access to "exotic" sexual practices that have been culturally coded as dangerous and risky.

Theorising the eRotics of HIV

In *Unlimited Intimacy* (2009), Tim Dean argues that gay men engaging in condomless anal sex are conceptualised as inhabiting a specific identity *as barebackers* and are theorised as engaging in a sense of shared subcultural community, a 'serofraternity' based on shared sexual practices that results in a common viral biology. Drawing on Kath Weston's influential research on queer kinship networks, Dean suggests that one of the central elements within 'breeding culture' is the reproduction not only of the virus, but of the virus *as a means of creating community*. Locating the emergence of barebackers as a distinct subculture in the 'post-AIDS' cultural context of the mid-to-late 1990s, Dean notes that the more conservative social agenda in gay politics that was necessitated by the AIDS crisis also facilitated a gay political milieu focused on gaining endorsement, both legal and cultural, of the extension of the heterosexual kinship paradigm to gays and lesbians; a discourse of gay respectability defined by the sociosexual norms of the heterosexual mainstream. In this sense, barebackers are understood as producing transgressive practices and identities 'to help keep their sex outside of the pale of bourgeois respectability' (2009, p. 85). For Dean, the 'bugchasing'/'gift-giving' dynamic of breeding culture in particular can be prismatically rendered in terms of fecundity and of a reinscription of how queer kinship can come to be.

'Bug chasers believe that being HIV positive makes them part of a gang', Dean argues, 'countering the image of the person with AIDS as an isolated outcast, voluntary seroconversion has come to be understood as a new basis for community formation' (p. 78). However, the emergence of new pharmaceutical interventions such as TasP, PEP, and PrEP has complicated the conceptualisation of barebackers as inhabiting a transgressive sexual identity based on proximity to risk and seroconversion.

This second antiretroviral revolution encapsulates a shift away from notions of viral fraternity and towards a neoliberal sexual individualism in which access to chemoprophylaxis enables immunity, both in terms of biological immunity and in the classical sense of the term as an exclusion from the obligations of the polity. '#TeenageTruvadaWhore' illustrates the ways in which eRotics permeate screen-mediated fiction. The scene reconstitutes the cultural value of gay pornography by explicitly centring HIV in the narrative. In doing so, director Liam Cole is producing a cultural intervention in the sexual and bio-politics of HIV, reconfiguring the cultural narrative of HIV/AIDS temporality through the prism of serofraternity. In '#TeenageTruvadaWhore', we see the young performer penetrated without condoms by seven older men, some donning viral inscriptions in the flesh by means of biohazard and Treasure Island Media tattoos intended to advertise positive HIV status. Yet in contrast to some of the other work of the Treasure Island Media studio in which "breeding" and associated semiotic forms that point to viral fraternity and viral reproduction are emphasised, here it is the new chemoprophylactic of Truvada and its potential for the user to "safely" transgress conventions of "risky" sexual behaviour that is in focus. The scene raises the question: if there is no virus being transmitted, can it still be called 'breeding'?

The concept of eRotics situates PrEP as a biomedical intervention with the potential of recalibrating notions of risk in embodied practices. As a pharmaceutical product with multivalent mobilisations—antiretroviral therapy for HIV-positive people; post-exposure prophylaxis for HIV-negative people; pre-exposure prophylaxis—Truvada can be understood as 'informed matter' (Rosengarten 2009) which is constituted not merely through its material properties but through the ways in which it is culturally articulated and socially incorporated into a broader assemblage of techniques of risk mediation. In the context of gay male sexual culture in the Global North, the utilisation of antiretroviral therapy in this way illustrates how 'biotechnologies are evidently active in what it is to act as a gay man in relation to the virus' (2009, p. 80). However, the geocorpographies of HIV/AIDS are rigidly bound by the contemporary logic of neoliberal globalisation, and the prospect of PrEP as a widespread risk mitigation strategy among HIV-negative gay men in wealthy countries must be juxtaposed against continuing struggles to ensure basic antiretroviral access for HIV-positive people in the Global South. In this sense, and in light of the discourses surrounding Truvada use discussed earlier, the same pharmaceutical compound represents a site of contest between those seeking erotic pleasure and those seeking life-saving treatment.

Geocorpographies of HIV

Joseph Pugliese coined the term 'geocorpographies' as a means of encapsulating the idea that 'the body, in any of its manifestations, is always geopolitically situated and graphically inscribed by signs, discourses' and 'regimes of visuality', and its 'geopolitical markings can only be abstracted through a process of symbolic and political violence' (2007, p. 12). I contend that PrEP is illustrative of the ways in which the seropositive body and the body "at-risk" of seroconversion can be read as a geocorpography. In this sense, it is not the virus itself which can be understood as *productive*, but the virus as mediated through antiretrovirals and articulated through neoliberal geopolitics. As such, inequitable global HIV health constitutes a pertinent example of what Michael Shapiro calls 'violent cartographies' (1997). That is, 'historically developed, socially embedded interpretations of identity and space' (1997, p. ix) which are 'constituted as inter-articulations of geographic imaginaries and antagonisms, based on models of identity-difference' (Shapiro 2007, pp. 293–294). The deployment of PrEP in the Global North engenders a semiotic shift in the cultural meaning of 'barebacking' from an identitarian practice functioning around a 'community' to an embodied immunitarian logic, demonstrating how both HIV and antiretrovirals are culturally shaped. By addressing the ways in which HIV as 'informed matter' also contributes towards the logic of the border—both the border of the body and the border of the state—the geocorpographies of HIV are revealed as geopolitical distinctions between similarly viral bodies. Three geocorpographic acts exemplify the ways in which the virus and the body are circumscribed forms of 'cartographic violence': the biometric gaze which restricts the movement of seropositive bodies, the geopolitics of the molecular mapping of HIV, and the apprehension towards providing widespread antiretroviral access to seropositives in the Global South due to concerns of 'nonadherence' to treatment regimens.

The biometric gaze is evidenced by the varying tests that seek to make a subject biometrically intelligible: hopeful migrants to Australia are subject to specific tests depending on the visa for which they are applying, their age, and intended occupation once settled. All migrants over 15 years of age who are applying for a permanent or provisional visa are subject to a HIV antibody test; if intending to work in the health sector, additional hepatitis B and C tests are required; if applying for an offshore humanitarian or protection visa, syphilis testing is also required (Department of Immigration and Border Protection 2015). Whilst Australian immigration law requires applicants to satisfy a 'health requirement', which involves screening for conditions that pose a 'threat to public health' or will incur 'significant cost' to the Australian community, the only specifically mandated biometric that is required for the health requirement is a HIV antibody test and a chest x-ray to check for tuberculosis (Australian Federation of AIDS Organisations 2013). For temporary visa applicants, the country of origin maps to a 'risk level' which determines the depth of the biometric gaze: temporary visa applicants from 'lower risk' countries—including the US, most of Northern and Western Europe, and some countries in the Middle East—are not required to undergo any health examinations in order to enter Australia, whereas 'higher risk' countries, including most of Eastern Europe, Sub-Saharan Africa, East and Southeast Asia, and Central Asia are subject to medical examinations for stays longer than 3 months (Department of Immigration and Border Protection 2015). Discussing the biometric (il)legibility of the bodies of refugees and asylum seekers in the European Union, Pugliese notes that bodies can be rendered illegible to the biometric gaze of the Eurodac fingerprint scan through a self-inflicted somatechnic intervention. He suggests that the cut of the blade across the flesh is a somatechnical act that enables the reconfiguring of the epidermis into that obtuse figure of the scar, disordered knitting of the skin that obliterates its own typologies and topographies. The refugee's act of self-mutilation mobilises this asignifying residue of flesh in order to fold the body upon itself and produce a moment of scarring occlusion and contingent non-meaning. (2010, p. 162)

This self-inflicted somatechnic act renders the subject illegible to the biometric gaze, creating 'minor and transitory "blockages" within the everyday operations of biometric systems' (p. 163) just as the hands of the Palestinian labourer, whose signifying typologies and topographies of the epidermis, eroded though manual labour, preclude the biometric gaze from identifying the body as a subject (p. 161). Whilst the biometric (il)legibility of these bodies produce temporary interventions into the 'foundational question of all biometric technologies' (p. 165)—that is, the question of 'who are you?'—the biometrics of HIV are concerned not with producing a coherent subject as a discrete, identifiable *person*, but rather attributing such a person with a quantifiable serostatus as a measure of allowing or precluding the movement of the subject across borders. Until 2010, seropositive people were unable to enter the US; presently, 13 countries prevent seropositive people from crossing the border, including Singapore, the United Arab Emirates, Russia, Papua New Guinea, and Iran (HIVTravel 2015). Many others employ categories of restriction to seropositive subjects: in Malaysia, migrant workers must demonstrate a negative HIV status in order to obtain a work permit (HIVTravel 2015).

Cartographies of biopolitical violence are not limited to individual bodies but are also enacted within the spatiality of the virus itself. Like the genome, the virus itself is mapped and represented spatially. The spatialisation and mapping of the various sub-types of the virus also produce geographic and economic inequalities for HIV treatment. Johanna Crane notes that 'virological and molecular biology research on HIV treatment have traditionally focussed almost entirely on the genetic "strain" or subtype of HIV found predominantly in North America, Europe and Australia' (2011, p. 144). The focus on the genetic subtype B of the virus as a 'reference strain' for research has meant that many antiretroviral (AVR) treatments that are successful in seropositives with this strain—most of whom live in the Global North—have been less successful when applied to non-B strains of the virus, which 'constitute nearly 90 per cent of the world's infections' (Spira, cited in Crane 2011, p. 144). As such, Crane argues that the geopolitics of HIV/AIDS, which already contribute to significant spatially-divergent experiences of HIV/AIDS by virtue of secondary factors of health outcomes such as economic and health inequality, are similarly present at the molecular level 'in laboratories where our knowledge about the molecular biology of HIV and ARVs is produced' (p. 145). In this sense, the seemingly objective task of 'mapping' the virus is revealed as '*productive* of certain possibilities' (p. 145). The seropositive body is thus a geocorpography insofar as it is corporeally inscribed by the virus and always already geopolitically situated because: (1) the determinants of HIV health outcomes overwhelmingly depend on a body's geographic location and political status as a subject, and (2) interventions into HIV are rigidly structured around the geopolitically situated molecular biology of the virus 'strain' selected for study.

The geocorpographies of HIV are plainly manifest in the apprehension towards antiretroviral therapy in the Global South. Whilst HAART proved remarkably successful in preventing the graduation of HIV into AIDS in industrialised nations following what Eric Rofes refers to as the 'Protease Moment' in 1996 'when all social and cultural changes in our experiences of the AIDS epidemic were explained in light of the new therapies' (1998, p. 29), calls to expand access to HAART in Sub-Saharan Africa were met with warnings. Harries et al. warned of 'antiretroviral anarchy' if HAART availability were to be 'widespread' and 'unregulated' in Sub-Saharan Africa, suggesting that the inability to monitor drug regimen adherence could lead to the 'rapid emergence of resistant viral strains' (2001, p. 410). Seropositives in the Global South were thus double-bound by the geocorpographies of HIV: being geopolitically situated in the Global South increased the secondary determinants that prevent ideal HIV treatment outcomes, which in turn constituted the justification for denying antiretroviral treatment in fear that an inability to adhere to the drug regimen would result in mutations of drug-resistant strains of the virus.

Conclusion: Viral Immunity

In contrast to the Global South seropositive body that is placed within a zone of antiretroviral exclusion, the "at-risk" seronegative body in the Global North is able to occupy a zone of viral exclusion. As such, PrEP constitutes the exemplary immunitary enabler. The Global North bare-backer, granted the proximity to eRotic pleasure, encapsulates both the contemporary epidemiological and the classical juridico-legal defini-tions of immunity. On the one hand, PrEP presents the potential for viral immunity, albeit through the daily incorporation of unnecessary toxic pharmaceuticals. On the other, this preventative regimen enables the subject to occupy an immunitarian serostatus, both HIV-negative and HIV-immune, and thus to abdicate obligatory requirements of gay male sexual hygiene as developed during the AIDS crisis. In the chemo-prophylactic, post-AIDS era, it is the notion of immunity as the incor-poration and control of the external threat that is most salient. The bifurcation of notions of viral immunity and the subsequent allowances of pleasurable behaviours and risky practices, for example, illustrates not only the ways in which immunisation relates to the incorporation of the threat into the body, but also in terms of immunity from the socio-sexual rules and consensus-driven norms of sexual contact in an age of HIV/AIDS. In other words, immunity can be understood in the ability for (mostly rich, mostly white, mostly inner-urban) gay men in the Global North to be prescribed PrEP and for condomless sex to be tolerated (if not endorsed) whilst many others, particularly those in the Global South, are treated as forms of bare life, with a seeming disinterest in providing equal or even adequate coverage for affordable antiretrovirals, especially if it might affect the income streams of mul-tinational pharmaceutical corporations. It is immunity that separates bare life from barebackers.

Note

1. Nostalgia here in the sense of a 'longing for a home that no longer exists or has never existed' (Boym 2001, p. xiii).

References

Adam, B. D. (2005). Constructing the neoliberal sexual actor: Responsibility and care of the self in the discourse of barebackers. *Culture, Health & Sexuality, 7*(4), 333–346.

Adam, B. D., Husbands, W., Murray, J., & Maxwell, J. (2005). AIDS optimism, condom fatigue, or self-esteem? Explaining unsafe sex among gay and bisexual men. *Journal of Sex Research, 42*(3), 238–248.

AIDS Council of New South Wales. (2015). *PrEP – Access options.* http://endinghiv.org.au/nsw/wp-content/uploads/2015/02/PrEP_Access_Options_Paper1.pdf#page=1. Accessed 2 Sept 2015.

Aronson, J. (1999). X marks the spot. *British Medical Journal, 318*(7197), 1543.

Australian Federation of AIDS Organisations. (2013). HIV and Australia's migration health requirement for permanent residence: Call for policy reform. *AFAO,* October. http://www.afao.org.au/library/topic/living-with-hiv/HIV-and-migration-position-paper_October-2013_FINAL_V3.pdf. Accessed 2 Sept 2015.

Bailey, A. (2008). Symbols: Historic and current uses. *International Journal of Pharmaceutical Compounding, 12*(6), 505–507.

Blechner, M. J. (2002). Intimacy, pleasure, risk and safety: Discussion of Cheuvront's "high-risk sexual behaviour in the treatment of HIV-negative patients". *Journal of Gay and Lesbian Psychotherapy, 6*(3), 27–33.

Boym, S. (2001). *The future of nostalgia.* New York: Basic Books.

Centers for Disease Control. (2011). Vital signs: HIV prevention through care and treatment – United States. *Morbidity and Mortality Weekly Report, 60*(47), 2 December, 1618–1623. http://www.cdc.gov/mmwr/pdf/wk/mm6047.pdf. Accessed 2 Sept 2015.

Centers for Disease Control. (2014a). *Preexposure prophylaxis for the prevention of HIV infection in the United States – 2014 clinical practice guideline.* Centers for Disease Control and US Public Health Service, 14 May. http://www.cdc.gov/hiv/pdf/PrEPguidelines2014.pdf. Accessed 2 Sept 2015.

Centers for Disease Control. (2014b). The Affordable Care Act helps people living with HIV/AIDS. *Centers for Disease Control and Prevention,* 2 June. http://www.cdc.gov/hiv/policies/aca.html. Accessed 2 Sept 2015.

Cole, L. (dir.). (2014). *London Uncut* [DVD]. Treasure Island Media.

Crane, J. T. (2011). Viral cartographies: Mapping the molecular politics of global HIV. *BioSocieties, 6*(2), 142–166.

Crary, D. (2014). Truvada, HIV prevention drug, divides gay community. *The Huffington Post*, 7 April. http://www.huffingtonpost.com/2014/04/07/truvada-gay-men-hiv_n_5102515.html. Accessed 2 Sept 2015.

Dean, T. (2008). Breeding culture: Barebacking, bugchasing, giftgiving. *The Massachusetts Review, 49*(1/2), 80–94.

Dean, T. (2009). *Unlimited intimacy: Reflections on the subculture of barebacking*. Durham/London: Duke University Press.

Department of Immigration and Border Protection. (2015). Health examinations. *Australian Government Department of Immigration and Border Protection*, 13 February. http://www.immi.gov.au/allforms/health-requirements/health-exam.htm. Accessed 2 Sept 2015.

Harries, A. D., Nyangulu, N. J., Kaluwa, O., & Salaniponi, F. M. (2001). Preventing antiretroviral anarchy in Sub-Saharan Africa. *The Lancet, 358*, 410–414.

Healy, P. (2014). A lion still roars, with gratitude. *The New York Times*, 25 May, p. AR1.

Heywood, T. (2014). The Truvada Wars: AHF launches misinformation campaign on PrEP. *HIV Plus Magazine*, 22 August. http://www.hivplusmag.com/prevention/2014/08/22/truvada-wars-ahf-launches-misinformation-campaign-prep. Accessed 2 Sept 2015.

HIVTravel. (2015). Global database on HIV related travel restrictions. http://www.hivtravel.org/. Accessed 2 Sept 2015.

Love, H. (2007). *Feeling backward: Loss and the politics of queer history*. Cambridge, MA: Harvard University Press.

McCasker, T. (2014). A porn director stirred up controversy by making a movie centered around HIV. *Vice*, 12 May. http://www.vice.com/read/director-paul-morris-believes-hiv-should-be-part-of-gay-porn. Accessed 2 Sept 2015.

McNeil Jr., D. G. (2014). Advocating pill, U.S. signals shift to prevent AIDS. *The New York Times*, 15 May, p. A1.

Morris, P. (1998). No limits: Necessary danger in male porn. http://www.managingdesire.org/nolimits.html. Accessed 2 Sept 2014.

Morris, P. (dir.). (2014). *Viral loads* [DVD]. Treasure Island Media.

Paasonen, S., & Morris, P. (2014). Risk and utopia: A dialogue on pornography. *GLQ, 20*(3), 215–239.

Pugliese, J. (2007). Geocorpographies of torture. *Critical Race and Whiteness Studies, 3*(1). http://www.acrawsa.org.au/files/ejournalfiles/65JosephPugliese.pdf. Accessed 2 Sept 2015.

Pugliese, J. (2010). *Biometrics: Bodies, technologies, biopolitics*. New York/London: Routledge.

Race, K. (2003). Revaluation of risk among gay men. *AIDS Education and Prevention, 15*(4), 369–381.

Race, K. (2009). *Pleasure consuming medicine.* Durham/London: Duke University Press.

Rofes, E. (1998). *Dry bones breathe: Gay men creating post-AIDS identities and cultures.* Binghamton/New York: The Haworth Press.

Rosengarten, M. (2009). *HIV interventions: Biomedicine and the traffic between information and flesh.* Seattle/London: University of Washington Press.

Shapiro, M. (1997). *Violent cartographies: Mapping cultures of war.* Minneapolis: University of Minnesota Press.

Shapiro, M. (2007). The new violent cartography. *Security Dialogue, 38*(3), 291–313.

Shernoff, M. (2006). *Without condoms: Unprotected sex, gay men and barebacking.* New York: Routledge.

3

Body, Crown, Territory: Geocorpographies of the British Monarchy and White Settler Sovereignty

Holly Randell-Moon

The British Monarchy operates through a system of geopolitical corporeality that requires an Anglican body to symbolise the sovereignty of divine rule. This body unites Commonwealth countries under a shared agreement to uphold and patriate Crown laws originating in the UK. Although Commonwealth countries are considered independent Realms under UK law, they nevertheless incorporate the Crown's religious authority into their secular laws. This chapter argues that the legitimation of Protestant theological rule through secular law is historically co-extensive with Crown possession and "discovery" of Indigenous lands in settler states such as Australia and Canada. That is, the secular autonomy of settler states is buttressed by Crown sovereignty. Drawing on Joseph Pugliese's term 'geocorpography' (2007), the chapter explains how the settler colonisation of Commonwealth nations reinforces the

H. Randell-Moon (✉)
Department of Media, Film and Communication, University of Otago, Dunedin, New Zealand

© The Author(s) 2016
H. Randell-Moon, R. Tippet (eds.), *Security, Race, Biopower*,
DOI 10.1057/978-1-137-55408-6_3

centrality of a white, Protestant body to sovereign forms of power in otherwise democratic polities committed to equality. I focus on recent changes to succession laws in the UK, which remove discriminating articles against females inheriting the throne and an heir's ability to marry a Catholic, to explain how whiteness and Protestant Christianity remain crucial to the geopolitical and corporeal exercise of Crown authority even as this authority is made compatible with liberal secular conceptions of "equal rights."

Settler states are colonised territories where a majority of white, British, or European migrants have demographically outnumbered the original inhabitants and assumed sovereign ownership of land through war and political exclusion of Indigenous peoples (Johnston and Lawson 2005, p. 361). The establishment of an autonomous parliament and judicial system is what demonstrates the ostensible termination of colonial ties between the settler state and its imperial place of origin. For instance, the Australian state came into being as a federated parliament in 1901, with a judiciary and legislative capacity that was distinct from the UK parliament, signalling the dissolution of colonial dependency. Central to such acts of independent statehood is the role of secular law in securing autonomous sovereignty. The Australian state is secular because its authority is derived from its own laws. What makes the law of a state sovereign, the highest authority within that state, is its secularity. Secular law, and the characterisation of this law as neutral and universal, is a crucial feature of the juridical affirmation of settler sovereignty and the extinguishment of Indigenous sovereignties as competing forms of authority (Randell-Moon 2013). Settlers justify their possession of territory through a seemingly neutral law that negates other sovereignties in order to preserve the autonomy of the settler state.

In theories of sovereign political power, the modern nation-state's sovereign ability to make laws through a secular parliamentary legislature replaces the theologically informed sovereignty of royal rule embodied in a monarchy. In the collection of lectures titled 'Society Must be Defended', Michel Foucault outlines how 'the rules of right that formally delineate power' (2003, p. 24) shift from the justification of royal power to the political power exercised by the mechanisms of the state: 'The juridical edifice of our societies was elaborated at the demand of royal power, as well as for its benefit, and in order to serve as its instrument or justification. In the West, right is the right of the royal command' (p. 25). The 'truth-effects' of law were then centred on the right of divine rule as a just and necessary exercise of power (p. 24). As parliamentary democracy became more secure, its legislature increasingly circumscribed royal command by reframing the legitimacy of royal power not just in terms of divine imprimatur (i.e. that a monarch need no other justification for rule other than god's blessing) but its compatibility and therefore limits within 'a juridical armature' of the state (p. 26).

Because early historical accounts of juridical rights were created to legitimise royal rule, Foucault argues that the 'problem of sovereignty' (p. 26) has dominated analyses of power. Such a focus draws attention away from 'local systems of subjugation' (p. 34) that take place outside of 'a single center' of dominance (p. 27). Nevertheless, the edifice and language of sovereignty still function as a legitimising force that organises and generates social conduct and relationships within state institutions. Where the sovereign power of the monarch was vitalised through the absolute submission of subjects to his rule, the emergence of state apparatuses in the eighteenth and nineteenth centuries relied on governmental techniques of subjugation centred on the preservation of life (Foucault 1991). Foucault terms this management of life 'biopower' and argues that racism and the delineation of aleatory bodies within the population became the sovereign prerogative of state and governmental institutions (2003, p. 255). He likens the 'dispositions, maneuvers, tactics, techniques, functionings' (1979, p. 26) used to foster governance of the population to a form of warfare over the rights to life. Here individual freedoms are not a right *to* something but are the means through which obligation and *submission* to institutions of power are carried out. For Foucault, 'politics is war by other means' (Foucault 2003, p. 15), indicating that the social body is constituted by a struggle over the discourses or truth-effects of power that render subjects governable.

Within this genealogy of rights and juridical power, it is the limitations or separation of royal and religious forms of divine right from (secular) state power that enables modern forms of political sovereignty and new techniques for governing the population. In the context of settler colonial states, however, the task of legitimising sovereignty persists because the existence of Indigenous sovereignties contests the juridical authority of both Crown and state sovereignty. Aileen Moreton-Robinson argues that connecting Foucault's articulation of state racism to the organising politics of settler supremacy allows us to see 'how colonisation operates through sovereign right as a race war' (2009, p. 64). The operation of Indigenous rights in the Australian legal system is then not taken for granted as the attribution of freedoms to Indigenous subjects but is rather a method of securing compliance to state and governmental management. Media and political attention on the "failures" of Indigenous subjects to conform to the legal requirements of the social body works as a discursive and institutional form of warfare that affirms the 'virtue' and legitimacy of the settler colonial state and the Crown possession that underpins it (p. 64).[1] The governing of Indigenous lives serves to demonstrate how contemporary state powers, achieved through settler colonialism, are implicated in the endorsement of older forms of royal sovereignty and right.

The media and political framing of the British Monarchy as now consistent with 'modern' society (Office of the Prime Minister 2015), because women can inherit the throne and an heir can marry a Roman Catholic, can also be understood as a form of biopolitical warfare that organises the social body in such a way as to legitimate Crown sovereignty and the exercise of state powers. Royal subjects are legitimate rulers precisely because they are able to be governed under parliamentary democracy. This subjection to governance is how the sovereign right of royal rule is preserved. Such preservation is consistent with the settler patriation of Crown possession of Indigenous territories as "lawful" and "just", through which Indigenous peoples are then rendered governable, per Moreton-Robinson. Affirmation of Crown possession and succession in the domestic laws of settler states serves to entrench the Anglo-British religious and cultural values of divine rule within those states' parliamentary and juridical arrangements, signalling an ongoing settler project to defend occupation in moral and legal terms. My focus on the amendments to the succession laws illustrates the geocorpography of Crown rule as inextricably tied to the racialised and spatialised configurations of settler possession. The "gender equal" and "religiously tolerant" amendments attempt to recuperate the virtue of Crown rule and serve to legitimate the exercise of state powers as equitable and for the common wealth.

The chapter will explain the legitimising function of monarchy for settler state sovereignty by first showing how the gendered and religious requirements of property inheritance in British common law are analogous to Crown "discovery" and possession of Indigenous lands. From here, the chapter explains that although the recent changes to succession laws are presented as "modernising" the anachronistic trappings of royalty, the changes nevertheless leave intact the religious and racial right of sovereign rule. The chapter concludes by locating the patriation of succession laws by the independent Realms of Canada and Australia within the geocorpographic relations of Commonwealth settler states.

Blood and Money

The British Monarchy is the embodiment of the Crown whose sovereign duty is to constitutionally delegate authority to the British parliament and supply a point of allegiance for Commonwealth nations. Constitutional monarchy is enabled through a series of governmental and sovereign arrangements whereby the Crown is simultaneously given consent to rule by the British parliament and the Commonwealth Realms but exercises the prerogative to allow parliamentary democracy to take place. At a domestic level, after an election takes place, the incumbent Prime Minister is required to meet with the King or Queen and asked to form a government. In this way, a Prime Minister is 'appointed' through royal prerogative but the monarch is also 'constitutionally bound' to follow the advice and exercise of prerogative powers carried out by government Ministers (Ministry of Justice 2009, p. 7). Historically, the monarch was vested with these powers through divine rule (at the will of god) (see Broom et al. 1869, p. 285–286). Prerogative powers have now been transferred from the Crown to the Cabinet and executive branch of parliament. As constitutional scholar A. V. Dicey explains, 'The prerogative is the name of the remaining portion of the Crown's original authority' (Dicey 1915, p. 421) and 'Every act which the executive government can lawfully do without the authority of the Act of Parliament is done in virtue of this prerogative' (p. 421). Thus, the contemporary British Monarchy has limited political powers within secular law and parliamentary arrangements, in keeping with the juridical transference of sovereignty to the state.

However, the preservation of the Monarchy through these same systems is often justified on the basis of Crown's ability to ensure extrajuridical continuity and stability. In an interview coinciding with the Diamond Jubilee of Queen Elizabeth II, former British Prime Minister David Cameron had this to say:

> And I ask myself, would I rather have some elected political president when you've got this institution that brings the country and the Commonwealth together, that's above politics, that symbolises nationhood and unity of the Commonwealth, isn't that far better, with all the links back into history and greatness that it has, rather than some here-today-gone-tomorrow president? (cited in Williams 2012)

The Crown's institutional durability is valorised by Cameron precisely because of its ostensibly apolitical and ahistorical role in Commonwealth civic and public life. Although the civic role of the Crown is often abstracted into symbolic and transcendental terms by monarchist commentators (see Randell-Moon 2015), it is nevertheless a political and religious system that is inherently and intimately corporeal. Even in the conversion of royal prerogative to parliamentary legislative supremacy, which is commonly read as a devolution of royal power, Crown sovereignty persists because it can be executed and extended through companion bodies capable of exercising its sovereign right.

The imperial and colonial expansion of the British Empire was predicated on the theological and juridical ability of the Crown to spatially disperse its sovereignty through acts of "discovery" carried out by explorers and colonial officers. Robert J. Miller outlines instances of emissaries from England and France burying coins and plates in the territories now known as Canada and the United States in order to juridically implement possession on behalf of their respective Crowns (2010, p. 21). He explains these acts as satisfying the Doctrine of Discovery, a series of international legal codes and principles formulated from the sixteenth century onwards and which outlined the criteria for "discovering" and claiming possession of land (p. 19). The Doctrine initially emerged in Catholic papal bulls, which were later inflected with Protestant Christian principles that linked the conversion of Indigenous peoples to Christianity with the establishment of civilisation in discovered lands (p. 16). The manufacturing of possession through rituals such as flag planting or burying items was a geocorporeal act of transforming discovered foreign lands into domestic Crown property. Miller explains that English feudal forms of property ownership and transference took place through 'a ritual called livery of seisin' where transferral of ownership occurred by 'turning over some dirt with a shovel and handing a clod of dirt or a twig from the property to the new owner for the witnesses to observe' (p. 22). Although these rituals of discovery have their antecedents in medieval and feudal laws, the juridical ability for the Crown to hold 'exclusive possession of its territory' (Moreton-Robinson 2009, n. 7/p. 78) is still 'the original and controlling legal precedent for Indigenous rights and affairs in Australia, Canada, New Zealand, and the United States' (Miller 2010, p. 2).

The execution of discovery principles through property rituals was also based on English juridical and ontological definitions of a property-owning subject as male, a gendering that (until recently) underpinned the corporeal requirements of monarchy. Explaining the role of primogeniture, where the firstborn male child inherits property, Vernon Bogdanor notes that in Britain 'under common law, the Crown descends on the same basis as the inheritance of land' (cited in Corcos 2012, p. 1598). Although primogeniture was removed by the *Succession to the Crown Act 2013*, which I will discuss later, hereditary aristocracy is still gendered as titles of nobility can only be bequeathed to males (which is why there is a gender imbalance in the House of Lords, whose membership is made up of appointed and hereditary peers) (Corcos 2012, p. 1658). The androcentric capacity for heritability in British law is premised on religious notions of divine right that have evolved to privilege the Anglican aristocracy and their related property rights under common law. Heritability as a noble right is linked to biologically informed ideals of aristocratic pre-eminence, for example those "born to rule" through "blue blood," a sanguine indication that the nobility's flesh is pale enough from lack of manual industry to express the transparency of venation. Within European history, this construction of nobility as biologically born to rule is also racialised in that lighter skin is associated with "bluer" veins (Lacey 1983). If the noble body fails to reproduce heritability, common law can intervene to protect patrilineal succession by allowing maternal inheritance once 'all his male maternal ancestors and their descendants have failed' (*Inheritance Act 1833*, cited in Corcos 2012, p. 1654).

In the patriarchal familial structure of Crown rule, the female body's biological reproductive capabilities are what simultaneously secures royal lineage through the production of male heirs and threatens instability through patrilineal marriage practices that dictate the absorption of women into the nominal male dynasty. Christine Alice Corcos describes the potential for female rulers as costing the Crown 'money and blood' (p. 1618) because her male consort would inherit the throne, upending 'the original ruling family' (p. 1609). British Queens have assumed the throne after primogeniture of the family had been exhausted, but this has caused some historical and common law anomalies. The current Queen of England, Elizabeth II, bequeaths her Windsor name and house to her children while her husband, Prince Philip, is barred from holding or assuming any constitutional power. Where a male sovereign through marriage passes on a commensurate title to his female partner (who becomes Queen), in the reverse, the male partner 'has no constitutional position' (p. 1664). The male body invested with the political-legal reproductive power of heritability can be stripped of these rights through customary law to secure patrilineal Crown succession when a woman, paradoxically in an aberration with that same customary law, assumes power.

This androcentric conception of inheritance is linked to particular political ideals about the types of bodies required for the operation of civil society. In his *Second Treatise of Government*, John Locke outlines how men evolve from a 'state of nature' into civil and social beings through the transformation of 'nature' into 'property' (1980, p. 28). Because a male body is capable of generating civility and property, he is rewarded with rights through law. Thomas Hobbes' *Leviathan* outlines how these rights-bearing bodies are implicated in an inter-corporeal social contract where each part of the social body contributes to the functioning of the community as a whole (1886, p. 11). The king or sovereign is figured as the head of the body, various governing bodies such as the judiciary are the joints and nerves, those who enforce laws are hands, explorers are the body-politic's eyes, and money and goods circulate as blood (Salavastru

2014). The metaphorical constituents of this body-politic are assigned differing levels of corporeal criticality reflecting their differing levels of importance to the operation of civil society. If the body loses its head, it will cease to function. If the body loses its hands, the amputated body, though enfeebled, can still operate. Common law and Enlightenment conceptions of a male body as able to bear rights and own property invest this body with a reproductive heritability that secures the development of civil society against the state of nature consigned to non-Christians, Indigenous peoples, and women.

Bodies that are male, Christian, and European are designated within Enlightenment philosophy and English law as best able to reproduce the conditions of civil society. Crown succession and preservation are also predicated on similar androcentric and religious concepts of heritable bodies that are inflected with racialised associations between the pale skin of the aristocracy and whiteness. Although Crown rule is based on a juridical form of sovereign right that is exclusive and divine, this divinity is nevertheless corporealised and requires bodies to exercise and carry out Crown authority. As a result, the laws regulating Crown rule are implicated in the state racism identified by Foucault as 'separating out the groups that exist within a population' in order to create a biological caesura (2003, p. 255) that historically worked to preserve those bodies conceived as necessary to the legitimacy of the Crown, and later the state and its exercise of powers. The gendered, religious, and racial values of English law ostensibly ameliorate the corporeal contingencies of Crown sovereignty so that the best bodies are throned to secure and reproduce royal sanguinity. Corporeal expressions of sovereign right then form a flesh point for the conflicts surrounding the presence of the Monarchy in British and Commonwealth political and juridical life. It is through the bodies of the Monarchy that Crown rule is understood, supported, and preserved in particular ways.

Revitalising Monarchy

In 2011, the David Cameron-led ministry introduced plans to amend existing succession laws to remove primogeniture and the disqualification of marriage to a Catholic after securing agreement from the sixteen countries in attendance at the Commonwealth Heads of Government Meeting in Perth, Western Australia (*BBC News* 2011). The implementation of changes to succession laws was described as 'head-snapping' (Corcos 2012, p. 1588) given that there had been thirteen attempts since 1979 to change the laws; in the past decade there have been three private members' bills alone (Twomey 2011, p. 2). The Succession to the Crown Bill (2012) was fast-tracked, a process whereby the government can enact laws swiftly by using the 'power of legislative initiative and control of Parliamentary time to secure their passage' (UK Parliament 2009, para. 27). This expedience was necessitated by the 'announcement that the Duchess of Cambridge … [was] pregnant' and the government's belief that there was a 'general consensus that the law should be changed as soon as possible' (UK Parliament 2012, para. 17). This executive exercise of law-making on behalf of public consensus proved unnecessary given the later birth of Prince George.

The succession changes were widely mediated through a discourse of gender equality and religious tolerance. For example, *BBC News* reported, 'Girls equal in British throne succession' (2011), the *Guardian* framed the changes as 'Royal succession gender equality' (Watt 2011) and *ElleUK. com* declared: 'The crown just got gender equal. Finally' (Lyons Powell n.d.). Cameron explained,

> The idea that a younger son should become monarch instead of an elder daughter simply because he is a man, or that a future monarch can marry someone of any faith except a Catholic—this way of thinking is at odds with the modern countries that we have become. (cited in *BBC News* 2011)

Similarly, former Canadian Prime Minister Stephen Harper commented that the changes ensure 'the Crown remains a vital and modern institution' (Office of the Prime Minister 2015).

Although the *Succession to the Crown Act 2013* was framed through a discourse of liberal rights, in terms of gender equality and freedom of religion, the right to the Crown is not subject to human rights provisions in the UK or the European Union. This is because, as the government points out in their explanatory notes on the Bill, 'The right to the Crown is not a private right' but 'a public right to the office of head of state, which is governed by statute' (UK Parliament 2012, para. 49). They go on to explain that 'although there are historical links between kingship and the ownership of land and the Crown brings with it the right to property,' the Crown 'is a right to inherit not property but the office of head of state' (2012, para. 50). Inheriting the Crown is thus opened up to women and spouses of Catholics under the auspices of liberal equality but the imprimatur of the Crown actually functions as a sovereign right to assume office.

The divine right of the Crown to exist and possess property is glossed by the government as 'historical' and the accumulation of property through possession is framed as seemingly incidental to the Crown's role as head of state. As noted above, the presumptive right of Crown possession of Indigenous territories forms the juridical basis of settler states and is anything but historical. This presumptive right to territory is described by Steven T. Newcomb as a kind of 'magic' that evacuates First Nations' sovereign rights (2011, p. 594). In another context, Moreton-Robinson describes how Yorta Yorta sovereignty was vacated by the Australian High Court as a result of the plaintiff's failure to demonstrate 'continuous' possession: 'Under the law of patriarchal white sovereignty, when a thief steals someone's property ownership is not assumed or inferred as being ceded to the thief. To the contrary, the law preserves ownership and guarantees return of the property to the owner' (2004, para. 12). Moreton-Robinson shows how domestic secular law works to guarantee Crown assumption of territory, contrary to its own common law principles. The sovereignty of settler courts in the United States and Australia to determine Crown possession non-justiciable (not subject to adjudication or examination by the courts), and therefore invalidate Indigenous property rights and sovereign claims, are coextensive with British parliamentary laws that likewise seek to preserve the presence of Crown rule in secular, democratic, legal, and political arrangements without subjecting this rule itself to justiciability.

While an heir is no longer disqualified from assuming the throne if their spouse is Catholic, the *Act of Settlement 1700* still prohibits a Catholic from inheriting the throne (Section II). The presumption underpinning this legislation is that any child from marriage to a Catholic will not inherit that religion and hence remain able to succeed the throne. No other religious adherent is specifically prohibited; hypothetically enabling a non-Christian to assume the throne. This sectarian prohibition stems from denominational incompatibility with the constitutional duties of the monarch as Head of the Church of England and the historical exclusion of Catholic heirs from the throne (Corcos 2012, p. 1623). As with primogeniture, the contemporary valency of sectarian discrimination has been publicly and legally contested. In 2013, Bryan Teskey challenged the domestic justiciability of upholding religious exclusion in the Canadian *Succession to the Throne Act 2013*. The Court of Appeal for Ontario upheld a decision from the lower courts that succession laws are not contestable by the Canadian Charter of Rights and Freedoms. The Court argued that 'The rules of succession are a part of the fabric of the constitution of Canada and incorporated into it and therefore cannot be trumped or amended by the charter' (Jones 2014). The Court also judged that 'Mr. Teskey does not have any personal interest in the issue raised (other than being a member of the Roman Catholic faith)' (Jones 2014). There is a distinction drawn here between an individual's religious freedom, framed as a 'personal interest', and the sovereignty of Crown rule, which forms an incontestable part of 'the fabric' of Canadian constitutional law requiring transnational consistency. In challenging the right of Crown rule under domestic rights legislation, Teskey is confusing sovereign right with liberal right. Sovereign right is exclusive and non-justiciable, and as Moreton-Robinson points out, is able to utilise law to protect itself from threats to its own authority. That Crown rule operates within the law as just and right, does not mean it is subject to the "ordinary" laws of citizens. This is illustrated in the Court's pithy statement that Teskey 'raises a purely hypothetical issue which may never occur, namely a Roman Catholic Canadian in line for succession to the throne' (Jones 2014).

The sovereign right of monarchy was framed through the succession amendments as congruent with contemporary liberal conceptions of gender equality and religious tolerance. Discourses of liberal equality work to abstract the bodies of the Monarchy from their geopolitical and imperial histories of war, discovery, and settlement. In coining the term geocorpography, Pugliese emphasises 'the violent enmeshment of the flesh and blood of the body within the geopolitics of war, race and empire' (2007,

p. 1). The bodies of the Monarchy are likewise intimately implicated in the geocorpographic reproduction of Crown and settler state sovereignty. This is precisely why the right to the British throne is not subject to secular liberal law. The bodies of the British monarchy are not substitutable with the ordinary abstract citizen of common law. They are set "above" these citizens through a biological caesura that manifests itself in the contemporary occupation of the throne by white and heternormative bodies. As a result of the geocorpographic system of Crown rule, the racial, religious, and gendered privileges of Anglican aristocracy are rendered a seemingly necessary and continuous part of the fabric of both domestic English and settler Commonwealth law.

Capillarising Settlement

I have explained so far that the preservation of Crown succession is historically linked to religious and racial conceptions of heritability that have been reaffirmed in domestic Commonwealth law to uphold the non-justiciability of Crown rule and possession. In this final section, I want to unpack how this transnational affirmation of Crown rule is predicated on an agreement that activates or capillarises the domestic independence of the Realms through subjection to the Crown. As explained earlier, Foucault insists that power operates and should be examined as a capillary-like movement through the social body rather than being exercised by or traced back to a central force in a controlling manner (2003, p. 27). Crown rule I argue, capillarises secular law in settler states by providing an impetus for law-making that demonstrates autonomy, but an autonomy that is directed towards compliance with UK law. This juridical capacity for autonomy through subjection is illustrated in the position of Commonwealth countries vis-à-vis the UK. Commonwealth countries are 'autonomous Communities within the British Empire' and under the *Statute of Westminster 1931*, are 'united by a common allegiance to the Crown' (Twomey 2011, p. 5). These Realms are made up of former settler colonies, such as Australia and Canada, or nation-states that have opted to join the Commonwealth, such as Tuvalu. As Anne Twomey notes, the Crown is 'divisible in its sources of advice and exercise of powers' but 'indivisible in its monarch' (2011, p. 25). What this means is that each of the Realms has a distinct monarch within their own legislative frameworks that is nevertheless tied to the same body inhabiting the English throne. The *Statute of Westminster* has been described by UK Lord Chancellor, Jack Straw, as a 'solemn compact' with 'huge moral force' (cited in Twomey 2011, p. 4). Straw's comments indicate that while the juridical power of the Statute has been extinguished, due to the separation of UK law from the sovereign state law of the Realms, Commonwealth nations nevertheless adhere to a geopolitical agreement to preserve the consistency of Crown rule (as the discussion of the succession laws and the Canadian Charter revealed).

In order to satisfy both the demands for sovereign independence and deference to the Crown, succession laws must be patriated consistently with those of the UK. For example, in Australia, the British Parliament does not have legislative power over domestic legislation. As a result, the Australian *Succession to the Crown Act 2015* was enacted to recognise the validity of its UK counterpart. The former law serves to constitute the Australian state as an independent sovereign entity even as this autonomy is inextricably tied to British law and all legislation must receive Royal Assent, performed by the Crown's representative, the Governor-General. The Crown's relationship to the Realms capillarises domestic forms of sovereignty by providing opportunities for settler states to undertake autonomous acts of law-making. The negation of Indigenous sovereign claims are consistent with these acts of Crown preservation, since such claims are typically adjudicated through, and often function to affirm, the settler juridical structure bought about by Crown discovery.

The independence of the Realms is generated through the autonomous and democratic operation of parliamentary law. This autonomy is secular because the domestic laws within a nation-state are juridically inscribed with the highest legal authority. I have argued elsewhere that to characterise liberal democratic states as secular is to obfuscate their racial and religious origins in imperial and settler colonial violence (Randell-Moon 2013). The secular framing of law works to present the juridical operation of the nation-state as neutral and universally applicable. In this way, the state itself cannot be cast as racist or sectarian. Rather, it is governed through neutral and secular laws which intervene and mediate between individual subjects who are racist or religiously intolerant and have thus infringed on the rights of others. This obfuscation of the racist and religious values in law takes place through what I have termed a 'secular contract'. Citizens contract into a secular state where their freedom of religion is protected in exchange for submission to secular law. Drawing on the work of Charles W. Mills (1997), I argue that this exchange also requires an epistemic ignorance about the racial and religious foundations of secular settler states. This is how the state's obvious political and juridical intersections with a theology of Protestant divinity, in the form of the Monarchy, are construed juridically and politically as consistent with contemporary social mores and discourses of equal rights.

The secular contract is implicated in the legitimation of the British Monarchy because the sovereign right achieved through discovery, androcentric heritability, and sectarianism is made commensurate with liberal democratic notions of equality in order to reaffirm the goodness or 'white virtue' (Moreton-Robinson 2009) of the Monarchy. Presenting sovereign right as consistent with secular law removes from view its racial and religious origins and the colonial endeavours that provided the blood and money to corporealise it. The secular contract is implicated in the race wars of contemporary colonialisms because secular law is used by the nation-state to declare its own authority and neutral independence, which delegitimises Indigenous claims to sovereignty, while at the same time, the nation-state submits to Crown rule. Put differently, secular nation-states such as Canada and Australia recognise only those forms of sovereignty that legimitise settler law.

Crown rule is corporeal and therefore tenuous, requiring continual efforts at vitalisation. For Foucault, analyses of power ought to move away from a Hobbesian model that conceives of power emanating from the 'head' or 'heart' of the social body. The peripheral limbs that capillarise this body are key sites of contestation and subjugation that constitute modern domains of power (2003, p. 29). I would suggest that the subjection of state instruments derived from colonial and imperial conquest to contemporary human rights discourse and legal conventions is an attempt to remove the sovereign monarch as head, replace it with secular state sovereignty, and consign monarchy to the outer limbs of the social body. There, as an ostensibly apolitical and symbolic feature of the Commonwealth, the Monarchy is effectively governmentalised. In moments of paraesthesia, such as the public spectre of male rule being embarrassingly inconsistent with a firstborn female child, the preservation of Crown rule then capillarises and animates the social body and the colonial sovereign heart of settler states. In the requirement for juridical coherency of royal rule across different Realms, spaces, and times in the one inter-corporeal body, it is the Crown appendages of settler states that kindle sovereignty.

Conclusion

The sovereign right of monarchy was framed through the succession amendments as congruent with contemporary liberal conceptions of gender equality and religious tolerance. Discourses of liberal equality work to abstract the bodies of the Monarchy from their geopolitical and imperial histories of war, discovery, and settlement. Such bodies are intimately implicated in the geocorpographic reproduction of Crown sovereignty as having a presumptive right to territory under the Doctrine of Discovery. The non-justiciability of Crown possession is part of the same juridical logic that governs the types of bodies able to occupy the throne, while the right to the throne itself is not contestable. The bodies enthroned by law are given sustenance by the property and wealth generated through imperial warfare and the stolen territories of Indigenous peoples legalised through settler colonialism. The media and political framing of the Monarchy as compatible with the contemporary body-politic, because of amendments to the succession laws, suggests that it is through subjection to parliamentary sovereignty that royal citizens receive rights in common with other citizens. This framing denies the geocorpographical constitution of rights as instrinsically tied to the asymmetrical positioning of bodies in space based on their corporeal significations as gendered, raced, and religious. Recognising how rights are governed geocorpographically illustrates the ongoing settler colonial ties between Commonwealth states and their juridical capillarisation through the British Monarchy.

Note

1. Moreton-Robinson uses the example of the Northern Territory National Emergency Response (known as the "Intervention") to connect biopower, state racism, and governmentality to colonial forms of sovereignty, suggesting that localised forms of subjugation that take place on the 'periphery' of the social body are violently inscribed with the problem of sovereignty. In this collection, Jillian Kramer considers how private property and the 'house' are used by the Australian state to discipline and punish Indigenous subjects and negate Indigenous sovereignties during the Intervention.

References

BBC News. (2011). Girls equal in British throne succession. http://www.bbc.com/news/uk-15492607. Accessed 5 Sept 2015.

Broom, H., Blackstone, W., & Hadley, E. A. (1869). *Commentaries on the laws of England* (Vol. 1). London: W. Maxwell.

Corcos, C. A. (2012). From agnatic succession to absolute primogeniture: The shift to equal rights of succession to thrones and titles in the modern European constitutional monarchy. *Michigan State Law Review, 2012*(5), 1586–1669.

Dicey, A. V. (1915). *Introduction to the study of the law of the constitution* (8th ed.). London: Macmillan.

Foucault, M. (1979). *Discipline and punish: The birth of the prison*. New York: Vintage Books.

Foucault, M. (1991). Right of death and power over life. In P. Rabinow (Ed.), *The Foucault reader: An introduction to Foucault's thought*. New York: Penguin.

Foucault, M. (2003). *"Society must be defended": Lectures at the Collège de France, 1975–1976* (D. Macey, Trans.) M. Bertani & A. Fontana (Eds.). London: The Penguin Press.

Hobbes, T. (1886). *Leviathan, or, the matter, form, and power of a commonwealth, ecclesiastical and civil* (2nd ed.). London: George Routledge and Sons.

Johnston, A., & Lawson, A. (2005). Settler colonies. In H. Schwarz & S. Ray (Eds.), *A companion to postcolonial studies*. Carlton: Blackwell Publishing.

Jones, A. (2014). Royal succession law not subject to charter challenge: Court. *CTV Kitchener News*, 26 August. http://kitchener.ctvnews.ca/royal-succession-law-not-subject-to-charter-challenge-court-1.1977651. Accessed 5 Sept 2015.

Lacey, R. (1983). *Aristocrats*. London: British Broadcasting Corporation.

Locke, J. (1980). *Second treatise of government*. Indianapolis: Hackett Publishing Company.

Lyons Powell, H., (n.d.). Big news for women in the royal family. *ElleUK.com*, 27 March. http://www.elleuk.com/now-trending/law-change-women-in-the-royal-family-The-Succession-To-The-Crown-Act. Accessed 5 Sept 2015.

Miller, R. J. (2010). The doctrine of discovery. In R. J. Miller, J. Ruru, L. Behrendt, & T. Lindberg (Eds.), *Discovering indigenous lands: The doctrine of discovery in the English colonies*. Oxford: Oxford University Press.

Mills, C. W. (1997). *The racial contract*. New York: Cornell University Press.

Ministry of Justice. (2009). *Review of the executive royal prerogative powers: Final report* (The Governance of Britain). http://www.peerage.org/genealogy/royal-prerogative.pdf. Accessed 5 Sept 2015.

Moreton-Robinson, A. (2004). The possessive logic of patriarchal white sovereignty: The High Court and the Yorta Yorta decision. *Borderlands e-journal*, *3*(2). http://www.borderlands.net.au/vol3no2_2004/moreton_possessive.htm. Accessed 15 June 2008.

Moreton-Robinson, A. (2009). Imagining the good indigenous citizen: Race war and the pathology of patriarchal white sovereignty. *Cultural Studies Review*, *15*(2), 61–79.

Newcomb, S. T. (2011). The UN declaration on the rights of indigenous peoples and the paradigm of domination. *Griffith Law Review*, *20*(3), 578–607.

Office of the Prime Minister. (2015). Statement by Prime Minister Stephen Harper on Canada providing assent to amendments to rules governing the line of succession. *Prime Minister of Canada*, 26 March. http://pm.gc.ca/eng/news/2015/03/26/statement-prime-minister-stephen-harper-canada-providing-assent-amendments-rules. Accessed 8 April 2015.

Pugliese, J. (2007). Geocorpographies of torture. *Critical Race and Whiteness Studies*, *3*(1). http://www.acrawsa.org.au/. Accessed 23 Sept 2014.

Randell-Moon, H. (2013). The secular contract: Sovereignty, secularism and law in Australia. *Social Semiotics*, *23*(3), 352–367.

Randell-Moon, H. (2015). Australian secularism, whiteness and the British monarchy. In T. Stanley (Ed.), *Religion after secularization in Australia*. New York: Palgrave Macmillan.

Salavastru, A. (2014). The discourse of body politic in Thomas Hobbes' Leviathan. *Les Cahiers Psychologie Politique*, *24*. http://lodel.irevues.inist.fr/cahierspsychologiepolitique/index.php?id=2613. Accessed 5 Sept 2015.

Twomey, A. (2011). Changing the rules of succession to the throne. *Sydney Law School: Legal Studies Research Paper*, 11/71, October.

UK Parliament. (2009). *Constitution Committee – Fifteenth report Fast-track Legislation: Constitutional implications and safeguards*, Session 2008–09, 17 June. http://www.publications.parliament.uk/pa/ld200809/ldselect/ldconst/116/11602.htm. Accessed 5 Sept 2015.

UK Parliament. (2012). Succession to the Crown Bill: Explanatory notes. Session 2012–13, 13 December. http://www.publications.parliament.uk/pa/bills/cbill/2012-2013/0110/en/130110en.htm. Accessed 5 Sept 2015.

Watt, N. (2011). Royal succession gender equality approved by Commonwealth. *The Guardian*, 28 October. http://www.theguardian.com/uk/2011/oct/28/royal-succession-gender-equality-approved. Accessed 5 Sept 2015.

Williams, P. (2012). British PM reflects on monarchy, jubilee and olympics. *7.30*, 4 June. http://www.abc.net.au/7.30/content/2012/s3517949.htm. Accessed 5 Sept 2015.

Legislation

Succession to the Crown Act 2015 (Australia).
Succession to the Throne Act 2013 (Canada).
Act of Settlement 1700 (United Kingdom).
Statute of Westminster 1931 (United Kingdom).
Succession to the Crown Act 2013 (United Kingdom).
Succession to the Crown Bill 2012 (United Kingdom).

4

What Are You Doing Here? The Politics of Race and Belonging at the Airport

Sunshine M. Kamaloni

Today I arrived back in Australia after being away for about a month. The airport is poignantly one of those places where I feel acutely aware of my difference. My otherness and foreignness becomes the centre from which I navigate my sense of self through the space. Perhaps it's an unwritten rule, but at the airport, you pretty much get treated according to the body you have and the label in your passport that accompanies it. Even though there is a sense of pride in who I am and where I come from, there is also a sense of self-consciousness and trepidation as I queue to have my passport checked and stamped. Before heading over to customs, I make a trip to the ladies' room mostly to gather myself together and put on my armour and this time it is my brave 'I can take anything' facade which I feel I need in this space. As I leave the rest room, a white middle-aged woman almost runs into me. I move out of her way and mumble an apology. I notice that she recoils from me, glancing at me with disapproving eyes with her

S.M. Kamaloni (✉)
Communications and Media Studies,
Monash University, Melbourne, VIC, Australia

© The Author(s) 2016
H. Randell-Moon, R. Tippet (eds.), *Security, Race, Biopower*,
DOI 10.1057/978-1-137-55408-6_4

forehead folding into a scowl and her mouth twitching uneasily. I see the disgust written all over her face, and she quickly moves away from me. Naturally, my heart sinks at this display, but I walk out telling myself that her behaviour has got nothing to do with me. And for a little while, I believe it. Later, as I wait in another queue, I find myself standing next to a black man with a trolley full of luggage. There is enough distance between us to indicate we are sole travellers, but I see the customs officer look at the man and then at me, and then ask the man, 'Are you together?' I marvel at the question because if we were together, shouldn't I be standing beside him or close enough? Should we be together simply because we are both black?

The first time I ever set foot in an airport was in 1987. I was saying good-bye to my father who was travelling to London, in the United Kingdom, for a one year work programme. He was leaving behind a wife and five children. I was six years old. And even at that age, I remember feeling uncertain about the space. There was a power that was apparent even to me, even as a child. Everything was arranged systematically—the primly dressed airport officials who took my dad's paperwork and ushered him along through the long corridors to the other side, away from us; the way his suitcase was tagged and I watched as it floated away on the conveyor belt among many other suitcases; and the way we were ordered to stand in one particular spot because we were not allowed to cross a threshold to say goodbye to my father. I remember being struck by this as a child, by feelings of separateness, unfamiliarity, strangeness, and loss in the space. And to this day, many years later, the airport remains a source of anxiety for me. There is always a sense of excitement and terror that fills me whenever I walk through the entrance to an airport. This mixture of emotions is always grounded in uncertainty—uncertainty about the processes that lie ahead, of checking in and being successfully cleared through; uncertainty about my bags and about the journey itself. Do I have my ticket? Did I pack my toothbrush? Did I put my aerosols in a see-through plastic bag? Do I have my passport? This is a mental checklist I run through every single time, without fail. The anxiety begins at least three days before a trip and it is accentuated by the long drive to the airport. The arrival at the airport is itself confronting in the way it highlights the state of arrival and yet departure all at the same time. This anxiety comes from the fear of not knowing whether and when I will be stopped or singled out for extra security checks.

With this spatial and affective context in mind, this chapter considers one question: what is it like to be a black woman in the airport? I want to locate this discussion within Joseph Pugliese's (2007a) *geocorpographies* in order to highlight how my body and by extension, racialised bodies, are geopolitically positioned in spaces like airports. I will do this by analysing my own experience in the airport, paying close attention to the relationship between race, bodies, and space in order to extrapolate the corporeal-racial-technology nexus of security and surveillance mechanisms in airports. Pugliese coins the term *geocorpographies* in his pioneering work on the technology of surveillance, law, and terrorism 'to bring into focus the violent enmeshment of the flesh and blood of the body within the geopolitics of race, war and empire' (2007a, p. 1). This encapsulates, first of all, how the body is always geopolitically situated, and secondly, the conceptual merging of the corporeal body with geography. Bodies come to be positioned in spaces through a process of symbolic, historical, political, and cultural discourse. In asking the question I have proposed, I aim to unpack the politics of bodies in space by interrogating how the airport as a space produces racialised particularities of experience.

Airports as Space

David Pascoe (2001, p. 34) beautifully describes the airport as a 'national frontier on the outskirts of a major city in the middle of a country.' Although not located at the territorial limit, this national space is indeed a frontier. However, it is a national frontier that connects to international spaces and a grounded site that embodies mobility (Salter 2007). Airports have been characterised as transition spaces (Gottdiener 2001); non-places (Auge 2008); spaces of authority (Kellerman 2008); sites of surveillance (Lyon 2003); seminal spaces for discussions of modernity and postmodernity (Creswell 2006); and as symbols of mobility (Adey 2004a). Marked with such versatility, the airport is very much a contradictory space. It can be fun and exciting on the one hand but serious and controlling on the other. While there are particular areas of the airport where these different faces are evident, there is nevertheless an acknowledgement that the serious face of the airport can impose itself in the more relaxed areas of the space at any time. It is partly this quality which makes the space of the airport unstable.

I want to argue that the airport is also a corporeal space. The environment of the airport makes bodies highly visible through technologies of surveillance, screening, and validating. These processes as well as the space itself elicit numerous kinds of feelings, from fear, anxiety, worry, frustration, panic, loneliness, disgust, pain, sadness, and boredom to excitement, happiness, pleasure, and euphoria. In my opening remarks, I talk about feeling acutely aware and self-conscious of my body and the otherness attached to it. As a black woman travelling alone, there is a sense of fear and vulnerability that accompanies me on most of my travels, particularly going through institutionalised spaces like the airport where a hierarchical order is evident. Entering the airport space triggers confusing feelings regarding self and identity. Arriving at Melbourne's Tullamarine airport, on which the case study for this chapter is based, is often conflicting for me in that it leaves me feeling disembodied. In a place where national identity matters, I feel neither Zambian nor Australian.

Peter Adey (2008, p. 151) argues that it is not an accident that the airport produces these feelings because the management of how an airport feels is not an arbitrary phenomenon. He posits that the airport can be understood as a 'differentiated landscape of intensity' where the airport's very design in turn affects passengers' emotions and their approaches and orientations toward particular forms of docility and conformity (Adey 2004b, 2008). Debbie Lisle (2003) suggests that these feelings, which come through intimate experiences of the space, are sometimes intended, engineered, and foreseen by airport authorities. This production of affect demonstrates how the airport space is produced, arranged, and managed. For instance, architects of the 1930s, 1940s, and 1950s, heavily influenced by the modernist pioneer Le Corbusier, wanted to create airports that would have the power to emphasise the pleasures, sensations, and wonders of air travel so that passengers would find it exciting. However, the potential for flight and its affective awe contrasts with the ways movement through the space of the border is heavily controlled.

Borders

According to Adey (2004a) airports are symbols of mobility. They are indeed a fitting representation of a world in the process of globalisation. And in such a world, how that mobility is managed and experienced becomes significant (Burrell 2008, p. 354). Airports have become places where millions of people flow in, out, and within countries each year. John Torpey has proposed that nations have the right and authority over the legitimate mobility of their citizens or as he calls it 'legitimate means of movement' (2000, p. 1). Therefore, the need for states to ensure that every movement into their territory is legitimate has evolved into a complex process of surveillance and control. The border zones through which people must pass have thus become the focus of intensified surveillance, checks, and control (Adey 2004b). In a mobile world, borders as barriers to movement and mobility can be experienced as a time-consuming nuisance. In a perceived insecure, vulnerable, and fearful world, borders become a necessary safeguard against illegal and threatening entrants. Mobility in this latter instance is seen as a risk to the safety and ordered territories of spaces.

I am particularly interested in the idea of the border as a spatial and corporeal boundary. David Newman (2006) argues for an understanding of borders as a process, a bordering process to be exact, as opposed to 'the border' itself. The understanding of the border as a process moves away from the traditional conceptualisations of borders as physical and highly visible lines of separation between political, social, and economic spaces to encompassing how they affect our lives on a daily basis, and most significantly at the very micro, interpersonal level (Newman 2006, p. 144). In essence borders create and reflect difference, a demarcation between countries, geographical spaces, and social spaces but also between people—us and them, insiders and outsiders. What is produced henceforth is a criterion of inclusion and exclusion: who belongs and who does not. Thinking about the bordering process allows us to (re) imagine and reflect on this criterion of belonging and how it (re)creates difference in the affective and corporeal space of the airport.

Orvar Lofgren (2004) identifies national borders as personal boundaries broadly articulating how the personal and by extension the corporeal cannot be separated from the idea of the border. Newman (2006) argues that narratives about experiences at borders reveal the borders that surround us on a daily basis and how our movement is restricted at those borders because we do not belong on the other side. In the world of border controls, bodies become somatic passwords, implicated in technologies of surveillance. David Lyon (2008) makes this point in light of the shift to identity management as a form of security control in public spaces, particularly at airports. And the key tool for identity management is biometrics. Individuals' personal data and information come under scrupulous scrutiny at borders even as their physical bodies are checked, verified, and identified. The picture that emerges from surveillance studies is one of airports as dangerous spaces that need to be surveilled for and secured from strange, illegal bodies.

Bodies and Mobility

It is important to locate the body in the airport precisely because it is through the body as an entity coming into contact with the space (the border) that we can explore and flesh out the embodied dimension of everyday living and knowing. The goal is to see the intimate link between the space and the body and therein the intimate link between the global and the everyday. The airport in and by itself embodies a sense of the global. It is the entryway to a new country and in many respects represents the country in which it is situated. But the coming and going of aeroplanes, peoples, and goods evinces a space that is linked to many different other spaces and peoples outside the bounds of the country. Mountz and Hyndman (2006, p. 447) argue that to express the global through the intimate and the intimate through the global, the sites of border, home, and body need to be explored. They liken the interconnectedness of the corporeal with these sites to a kaleidoscope, with border, home, and body each blurring the global and the intimate into the 'fold of quotidian life' (p. 447). Hence we come to see how bodies are geopolitically positioned in the global through the intimate inter-corporeal encounters that constitute everyday lived space.

Mapping these intimate experiences onto space highlights the ways in which the meanings and uses of space have more to do with determining and assigning who does and who does not have the power to define, claim, and control space (Neely and Samura 2011, p. 1939). I want to interlace this with the concept of geocorpography precisely because bodies do not exist as abstract entities in a vacuum but rather are always already socioculturally positioned and graphically inscribed (Pugliese 2007b, p. 125). The emergence and function of corporeality within any given space is thus determined by the complex interplay of historical, political, and socio cultural factors (Pugliese 2007a). For non-white bodies, for instance, public spaces like the airport become spaces of anxiety and danger where they are subjected to repeated security checks and harassment and the possibility of both symbolic and physical violence (Pugliese 2006). These power relations are played out through racial interactions in spaces. This is why the relationship between space and race is defined by inequality and difference (Neely and Samura 2011; Rollock et al. 2011; Shields 2006). Edward Said's (1978) work on 'orientalism,' and particularly his concept of 'imagined geographies,' traces out the spatial dimensions of imperialism. He asserts that the meanings of race and space are constantly created and recreated in relation to an 'other.' This understanding makes more apparent the symbolic and material ways in which race becomes embedded into the spatiality of everyday social life. One of the key questions to keep in mind while examining my experiences at the airport is thus, what social relations and identities are being produced and reproduced through the airport as a geocorpography of social and physical space?

The Somatic Norm Versus All Other Bodies

When the customs officer asks me whether I am with the black man in front of me in the queue, my first reaction is surprise. I had never seen the man before in my life. He exists to me only in the moment I realise he is the person ahead of me. There is nothing in my behaviour that indicates that I know the man or am even acquainted with him. Therefore, my natural inclination is to question why the customs officer assumes that the "strange" black man and I are together. There is a tension between the way the officer reads us and defines us as a couple and the way we read ourselves as distinct black individuals. It is clear that the officer's reading of the situation is not based on any concrete, material evidence of couple-ship. This is in stark contrast to various traveller couples in the same queue who are standing very close to one another, talking to each other, and helping each other with their luggage. In our case, as I note in my opening narration, there is enough distance between us to let anyone know that we are simply two people standing in a queue. The officer's definition of us as a couple brings race into his surveillant perception: that we both look "the same" and so by implication we are or should be together. The officer draws from a racialised 'regime of visuality' that activates the stereotypical pairing and thereby resignifies my individual identity into a collective one (Pugliese 2006).

The officer's attempt to locate our bodies within the space involves a process that sees him linking separate bodies together and reading them as a unit—a representative whole. Pugliese (2005, p. 11) in his seminal work on biometrics notes the historical progression of a racialised visuality in surveillance in the inability of British colonial officers to read ethnic difference in places like India. British officials complained of the 'problem of racial homogeneity' characterised by similar physical features such as hair, eyes, and skin complexion. I refer to this geocorpography not only because it highlights the historical and socio cultural context of the airport official's surveilling gaze but because it also demonstrates the specificity of the manifestations of my racialised experience. Locating me as the black other allows the officer to make the assumption that I am connected to a random black man. In assuming this connection, the officer taps into general and historical beliefs about black people being an undifferentiated mass, a people whose unique individuality and positions become lost as they are decontextualised and instead viewed as interchangeable—one black person is the same as another black person (Collins 2000; hooks 1981).

While it is not clear whether the security officer finds the possibility of the black man and I "being a couple" threatening or reassuring, his assumption focuses my attention on his white gaze. This type of superficial collating of bodies not only perpetuates racial stereotypes—i.e. a black woman should be with a black man—but it also speaks to larger historical and political conditionings of reading the racialised body.

A benign reading of the officer's gaze is that his job simply requires him to group similar people together as a form of racialised risk profiling in a disorganised space. However, it is important to recognise this gaze is not merely an innocuous reflex of conditional training. It is, in fact political. The deployment of surveillance technologies is highly racialised, and, as Pugliese attests, it is another 'instantiation of unacknowledged whiteness' (2005, p. 2). Biometric systems, which are calibrated to whiteness and identify and verify subjects based on phenotypical norms, inscribe black bodies as problems for computation. Black bodies are more likely to fall outside the operating parameter of a biometric system because the technology precludes the biometric capture of the features of non-white subjects (Pugliese 2005, p. 5). This greatly influences how some bodies are rendered more visible or invisible. I want to emphasise here the contradiction between the white bodies in the queue and my own body as well as the black man's. The white bodies are not questioned, their identities speak for themselves. Whiteness as an ethnic particularity has always been unmarked and unseen for those in power. Therefore, the white body acts as the universal 'somatic norm' (Puwar 2004). Nirmal Puwar defines the somatic norm as 'the corporeal imagination of power as naturalised in the body of white, male/middle class bodies' (2001, p. 652). Bodies which are the somatic norm have an undisputed right to occupy a space (Puwar 2004, p. 8). They are the geocorpographies of a sustained colonial power evident in the practices of everyday life.

The juxtapositioning of white bodies and black bodies in a historically white space, such as the airport, demonstrates how particular bodies are rendered natural to the space. Sara Ahmed argues that 'white bodies are comfortable as they inhabit spaces that extend their shape' (2007, p. 158). These spaces in which white bodies are comfortable are the ones whose surfaces have already been impressed by white bodies through the familial repetition of inhabiting them. The juxtaposed encounters between white bodies and black bodies in seemingly ordinary spaces highlight their geocorpographic dimensions, that is, the constructed historical and imaginary boundaries assigned to particular bodies. Therefore, despite the legal right for all bodies with the correct documentation to enter the airport space, there exists a covert and subtle process of inclusion and exclusion which continues to casually operate through the differentiation of the somatic norm from all *other* bodies. This somatic differentiation underscores and is reinforced by the racialised regimes of security profiling and surveillance technologies in airports. As Ahmed (2007) contends, having the right passport does not make a difference if you have the wrong body. Ahmed (2004) has suggested that some bodies, more than others, are recognised as strangers and as bodies out of place.

This perceptual relegation of bodies to their "appropriate" spaces greatly impacts on the everyday location of racialised bodies, as is evidenced by the officer's reaction to my proximity to another black body. The officer's gaze draws attention to my invisibility as well as my visibility. I am invisible inasmuch as the officer fails to read the social cues that indicate that I am a woman travelling on my own. This is indeed troubling because his role as a security officer ostensibly demands acute and accurate perception. At the same time, my visibility is highlighted in the way I am only allowed to exist in that particular space at that particular moment as an identity that he perceives and judges appropriate for a black woman—attached to another black person. I feel that I am not being seen or acknowledged as an individual in my own right. Research has shown that black people overwhelmingly report white people's inability to recognise and distinguish black faces unless impressed upon them in a 'relationship' (Feagin and Sikes 1994). In these encounters, many black people describe feeling invisible—that their presence or contribution is judged to be less valuable (Sue et al. 2008, p. 334). The officer lumping me together with the unknown fellow traveller further differentiates our bodies from the other bodies in the queue, which happen to be predominantly white-skinned.

The Feeling Body

Standing in the airport queue, there is an acute awareness of my body: of my black skin and my differently textured hair, which is often hidden in braids. I am uncertain about how I will be perceived and processed. As a result, I am invaded and overwhelmed by a sense of anxiety and powerlessness at the lack of control over how I will be seen and therefore how I will be evaluated as a secure or risky body. I become alert and edgy, keeping my attention focused on everything that is happening around me. I notice the

way the airport officials interact with travellers of different races and nation-
alities. I notice the looks given, the nature of the questions asked and then
I brace myself for the worst. There is an internal dialogue constantly occur-
ring about whether I have the right to pass through the border and whether
I have committed any crimes I wasn't aware of that may deny me passage.
And suddenly I'm feeling unsure about my legality and guilty about things I
did not do. The awareness that there are forces beyond my control that have
the power to determine my identity within the space as well as my legality
manifests itself in feelings of anxiety and uneasiness within my body—heart
thumping, shaking, and holding my breath.

The corporeal and affective characteristics of my experience in the
airport are emblematic and a critical part of how I experience the world
as a black woman. The inner dialogue and the manifestations of anxiety
and fear in the physiological forms of increased heart rate, shaking, and
fast breathing is a constant in many of the racial experiences I have had.
This not only demonstrates the corporeal nature of racism and spaces
but also the importance of understanding the body as the centre of experience. However, these internalised responses often go unacknowledged
or critically unexamined in the moment of impact or in theories about
race. The reactions become something one has to hide—particularly if
they have to do with strong emotions like anger. For example, studies
reveal that there is a general perception from white people that the emotions of black people, especially anger, are not appropriate and should
therefore be repressed (Feagin and Sikes 1994; Wingfield 2010, p. 259).
This negative affective perception is responsible for the stereotypes of
black people as "angry."

As a black woman I am conscious of this representation, and to avoid
perpetuating the stereotype I have often suppressed my anger at the dif-
ferential treatment I receive in different spaces. bell hooks (1981, p. 86)
has argued that many black women avoid publicly expressing emotions of
bitterness or anger for fear of being labelled with the Sapphire identity—
a stereotypical identity that was historically projected onto any black
woman who overtly expressed anger, rage, or bitterness about her life
situation. The point I want to emphasise here is the way in which racial
incidents set me at war with myself. This is an important theme that sets
up a key problem: racism becomes a struggle with oneself because, as an

idea, colour-blindness essentially and practically hides the real, complex, and contextual experience of race in ordinary spaces.

The collision between the border and bodies is not the only type of encounter happening in the airport. The constant movement in the space means that bodies meet and clash with other bodies all the time. These are the types of encounters I am calling 'body on body' encounters, and they illustrate how the body acts as a border through abjection. Julia Kristeva in *Power of Horror* (1982) defines the 'abject' as the human reaction to a breakdown in meaning caused by the loss of distinction between the subject and object or between the self and the other. I am using abjection here to emphasise the movement across bodily borders related to different kinds of people coming into contact with each other. Abjection as a concept serves to reinforce or maintain the boundary between the self and the other (Kristeva 1982; Philips 2014).

The Body as a Border

The function of the national border is to protect against illegal entry or contamination, and the same concept applies to bodily borders. With the proximity of bodies in the airport there is a fear of contamination and disease. The body on body encounter is another form of encounter that happens in the airport from which we can draw narratives about the location of the racial body in space. My encounter with the woman in the restroom mentioned in the opening of this chapter is an example. It acts as a mirror reflecting the global onto the intimate. This is because the encounter felt like an extension of the larger airport processes happening at the passport control point, the customs check counter and the security check point made up of checking and rechecking, identification, and classification of bodies and the inter-corporeal anxiety that is embedded in these practices and through which they are sustained. The woman's first reaction to our near collision is surprise, then anger, and finally, disgust. My own reaction to the woman and our encounter is varied. Like her, I am surprised when she suddenly appears behind the door as I attempt to push it open. I do not expect that and thus, her sudden presence catches me off guard. I offer an apology as I step aside to allow her to walk through first. I make eye contact seeking her acknowledgement of my apology. Instead, I see confusion, disapproval, and revulsion on her face. This initially throws me and I have my own visceral response to her reaction. My first thought is about my skin and I become acutely aware of the blackness of it. I wonder if she is associating it with dirt, bad smells, and uncleanness epitomised by the very space within which we stand. As my heart beats faster at the thoughts, I am flooded with embarrassment.

Jackie Stacey (1997, p. 82) argues that bodily revulsion is the sign of abjection, the abject being that which needs to be expelled and excluded (Kristeva 1982). According to Newman (2006), animosity and dislike for the other takes on a concrete form through the act of meeting. So when the woman and I collide in the bathroom, what in the past may have been only the woman's awareness of the other based on invisibility and a lack of knowledge now takes on a material and definite form. I become the materialised abjection. When it comes to bodies, abjection, or the loathing of other bodies is political in the way it plays a crucial role in the ritual exclusion of people from what is considered in need of protection or preservation (Stacey 1997, p. 75).

According to Mary Douglas (1966, p. 2), rituals of cleanliness and purification are symbolic of ordering and organising the environment. Therefore, dirt and by extension "dirty" bodies are a sign of disorder. The airport is particularly interesting in this regard as it requires people to be sanitised. People coming from particular countries are asked questions about certain diseases, their health during their stay in foreign countries, and whether they have been immunised or sanitised. Any threat is punished through detention or denial of the right to cross the border. There is fear of and anxiety regarding the introduction of pandemics, diseases, and bodily pollution which may create disorder. Therefore, the airport is bound up in the politics of regulating and monitoring not only the nation's borders but the borders of bodies as well.[1] In relation to these anxieties of contamination, pandemics, and pollution, bodies have been understood as organised within and through a system of boundaries, which are fundamental to beliefs about contamination and sanitisation (Stacey 1997, p. 75).

The threat posed by the bodies of others to bodily and social norms is registered on the skin or as Ahmed eloquently puts it, 'the skin comes to be felt as a border through the violence of the impression of one surface upon another' (2004, p. 54). Thus, I argue that the white woman experiences this violence when my black body unexpectedly impresses upon the surface of her own white skin, causing her to react to the collective consciousness and representation of blackness as opposed to whiteness. Her reaction collects into one final emotion which displays before she turns away from me—disgust. Disgust, like many other emotions, involves the emergence of bodies when we encounter others in intimate and public spaces (Ahmed 2004, p. 55). As with hatred, disgust is a negative bodily and affective connection to another that one wishes to be removed from, a connection that Ahmed argues is sustained through the removal or expulsion of the other from bodily and social proximity. The woman's disgust and her reaction of moving quickly away from me indicate feelings that her physical or personal space has been invaded. She moves hastily aside, as though my presence is engulfing her, illustrating that even one single non-white body can be seen to be taking up more physical space than it actually occupies. Intertwined with this imagined idea of racial amplification is the phenomenon of visibility, as bodies considered out of place, or unexpected bodies, become highly conspicuous. Sometimes this results in violence and fatality, as in the case of the police shooting of Jean Charles de Menezes, who was racially profiled—his Brazilian ethnicity resignified to South Asian—and in the process criminalised, condemned, and killed (Pugliese 2006).

Puwar (2004, p. 49) suggests that this racial visibility comes from non-white bodies not being the somatic norm. The amplification happens precisely because the geocorpographies of non-white bodies are known and perceived in ways that are seen to threaten the artificial claims to space for the dominant, superior, and normalised white identity. There is an evident anxiety on the part of governing bodies, a fear—terror almost—of numbers, a fear of being swamped and taken over. And this necessitates management and control on behalf of those bodies that are ascribed racial normality and safety. In other words, this amplification reveals the geocorpographic underpinning of bodies in airports and the technologies of security used to contain somatic risk.

Conclusion

The airport as a space can be seen as a reflection of larger geocorpographic concerns—a corporeal map of how bodies are geopolitically positioned in space and a model for organising cities into heavily surveilled, heavily controlled, and managed spaces (Sorkin 2003, pp. 261–62). This is especially so in our post-9/11 environment in which there is a faith in new technologies (particularly identity securitisation and management technologies which are often first tested at airports) to solve social and political problems, particularly in North America (Lyon 2008, p. 30). The heavily surveilled airport is a creation of this corporeal-racial-technology nexus. The geocorpography of the modern international airport must be understood as ultimately a contact zone through which certain bodies are rendered invisible as subjects but visible as objects. The airport highlights the power imbalance between bodies, even as it further comments on levels of fear and anxiety in an increasingly globalised world—the fear and anxiety exhibited at the airport is particularly over the borders of bodies, bodies on to which illegality, disease, and foreignness are mapped. And this, I argue, is indicative of larger social issues of race and managing the other. As a lived space, the airport is a model of Western society itself. It is within this context that I can answer the question I began in the introduction, and make sense of the affective dimension of my experience as a black woman in the airport. I am subject to a racialised visual regime of surveillance that not only positions me as a threat in the space but also sets me at war with myself. The geocorpography of my black body is at once troubling and exhausting as I navigate the corporeality of everyday life.

Note

1. See Joshua Pocius this volume for a discussion of the viral boundaries and borders of bodies in the context of HIV/AIDS pharmaceutical intervention and prevention.

References

Adey, P. (2004a). Secured and sorted mobilities: Examples from the airport. *Surveillance and Society, 1*(4), 500–519.

Adey, P. (2004b). Surveillance at the airport: Surveilling mobility/mobilising surveillance. *Environment & Planning A, 36*(8), 1365–1380.

Adey, P. (2008). Mobilities and modulations, the airport as a difference machine. In M. B. Salter (Ed.), *Politics at the airport*. Minneapolis/London: University of Minnesota Press.

Ahmed, S. (2004). *The cultural politics of emotion*. Edinburgh: Edinburgh University Press.

Ahmed, S. (2007). A phenomenology of whiteness. *Feminist Theory, 8*(2), 149–168.

Auge, M. (2008). *Non-places: Introduction to an anthropology of supermodernity* (2nd ed.). London/New York: Verso.

Burrell, K. (2008). Materialising the border: Spaces of mobility and material culture in migration from post-socialist Poland. *Mobilities, 3*(3), 353–373.

Collins, P. H. (2000). *Black feminist thought* (2nd ed.). New York/London: Routledge.

Cresswell, T. (2006). *On the move: Mobility in the modern western world*. New York: Routledge.

Douglas, M. (1966). *Purity and danger: An analysis of concepts of pollution and taboo*. London: Routledge/K. Paul.

Feagin, J., & Sikes, M. (1994). *Living with racism*. Boston: Beacon.

Gottdiener, M. (2001). *Life in the air: Surviving the culture of air travel*. London: Rowman and Littlefield.

hooks, b. (1981). *Ain't I a woman: Black women and feminism*. Boston: South End Press.

Kellerman, A. (2008). International airports: Passengers in an environment of "authorities". *Mobilities, 3*(1), 161–178.

Kristeva, J. (1982). *Powers of horror: An essay on abjection* (L. S. Roudiez, trans.). New York: Columbia University Press.

Lisle, D. (2003). Site specific: Medi(t)ations at the airport. In F. Debrix & C. Weber (Eds.), *Rituals of mediation: International politics and social meaning*. Minneapolis: University of Minnesota Press.

Lofgren, O. (2004). My life as a consumer. In M. Chamberlain & P. Thompson (Eds.), *Narrative & genre – Contexts and types of communication*. New Brunswick/London: Transaction Publishers.

Lyon, D. (2003). Airports as data filters: Converging surveillance systems after September 11th. *Information, Communication and Ethics in Society, 1*(1), 13–20.

Lyon, D. (2008). Filtering flows, friends, and foes: Global surveillance. In M. B. Salter (Ed.), *Politics at the airport*. Minneapolis/London: University of Minnesota Press.

Mountz, A., & Hyndman, J. (2006). Feminist approaches to the global intimate. *Women's Studies Quarterly, 34*(1/2), 446–463.

Neely, B., & Samura, M. (2011). Social geographies of race: Connecting race and space. *Ethnic and Racial Studies, 34*(11), 1933–1952.

Newman, D. (2006). The lines that continue to separate us: Borders in our "borderless" world. *Progress in Human Geography, 30*(2), 143–161.

Pascoe, D. (2001). *Airspaces*. London: Reaktion.

Philips, R. (2014). Abjection. *TSQ: Transgender Studies Quarterly, 1*(1–2), 19–21.

Pugliese, J. (2005). In *silico* race and the heteronomy of biometric proxies: Biometrics in the context of civilian life, border security and counterterrorism laws. *Australian Feminist Law Journal, 23*(1), 1–32.

Pugliese, J. (2006). Asymmetries of terror: Visual regimes of racial profiling and the shooting of jean Charles de Menezes in the context of the war in Iraq. *borderlands e-journal, 5*(1). http://www.borderlands.net.au/vol5no1_2006/pugliese.htm. Accessed 3 Oct 2015.

Pugliese, J. (2007a). Geocorpographies of torture. *Critical Race and Whiteness Studies, 3*(1), 1–18. http://www.acrawsa.org.au/files/ejournalfiles/65JosephPugliese.pdf. Accessed 2 Sept 2015.

Pugliese, J. (2007b). Biometrics, infrastructural whiteness, and the racialised zero degree of nonrepresentation. *Boundary 2, 34*(2), 105–133.

Puwar, N. (2001). The racialised somatic norm and the senior civil service. *Sociology, 35*(3), 651–670.

Puwar, N. (2004). *Space invaders, race, gender and bodies out of place*. Oxford/New York: Berg.

Rollock, N., Gillborn, D., Vincent, C., & Ball, S. (2011). The public identities of the black middle classes: Managing race in public spaces. *Sociology, 45*(6), 1078–1093.

Said, E. W. (1978). *Orientalism*. New York: Pantheon Books.

Salter, M. B. (2007). Governmentalities of an airport: Heterotopia and confession. *International Political Sociology, 1*(1), 49–66.

Shields, R. (2006). Knowing space. *Theory, Culture & Society, 23*(2–3), 147–149.

Sorkin, M. (2003). Urban warfare: A tour of the battlefield. In S. Graham (Ed.), *Cities, war and terrorism: Towards an urban geopolitics*. Oxford: Blackwell.

Stacey, J. (1997). *Teratologies: A cultural study of cancer*. London/New York: Routledge.

Sue, D. W., Nadal, K. L., Capodilupo, C. M., Lin, A. I., Torino, G. C., & Rivera, D. P. (2008). Racial micro-aggressions against black Americans: Implications for counselling. *Journal of Counselling and Development, 86*(3), 330–338.

Torpey, J. (2000). *The invention of the passport: Surveillance, citizenship and the state*. Cambridge: Cambridge University Press.

Wingfield, A. H. (2010). Are some emotions marked "whites only"? Racialised feeling rules in professional workplaces. *Social Problems, 57*(2), 251–268.

Part II

Technologies

While we have established so far in this collection that "technology" can be a very broad concept referring not only to technical, but also social formations, the three chapters in this section deal with the kinds of "technology" which that term might traditionally call to mind. In particular, all three contributions deal with and problematise different aspects of digital media, highlighting the tangible, the corporeal, and the material impacts of technologies and processes that define and market themselves as "weightless", "green", and "gaseous".

In Ryan Tippet's chapter, Facebook's Internet.org program to institute 'zero-rating' internet, where users have unlimited free access to a selection of services, comes under scrutiny. Brett Nicholls turns the focus in Chap. 6 towards those wearable technologies that track and measure individual biometric data, in order to instruct and encourage subjects towards health and fitness betterment. Lastly, Sy Taffel visits four moments in the life cycle of digital culture gadgetry—from Rare Earth Element mining, to assembly lines in exploitative outsourcings, through precisely micromanaged Amazon warehouses, and finally to toxic and deleterious e-waste sites in Africa and Asia.

These chapters, in their focus on technology, do not lose sight of race or somatechnics. Internet.org is conceived, ultimately, as 'corporate geocorpography'—a racialised production of bodies in certain regions as at once "lacking" and an untapped commercial resource. Wearable health motivation technologies aim to manipulate and modulate the very bodies of their users in the nexus of biopolitical governmentality and somatechnics, rendering bodies and processes as so many normative, aspirational, manipulable data points in a competitive neoliberal modulation. The artefacts of digital ecology affect many people within their life cycles, but their overlooked or concealed human and environmental costs are by no means evenly distributed across class and race divisions: what privileged consumers might experience as "weightless" technology may contribute, at the end of its limited use, to cancer and nerve damage in the "artisanal" e-waste workers of the Global South.

A familiar theoretical thread runs throughout these chapters. Gilles Deleuze's (1992) brief summary of the 'Control Society' future of Foucauldian power is an almost-obligatory reference in contemporary accounts of digital technology. Deleuze believed that the disciplinary institutions described by Foucault 'are the history of what we are slowly ceasing to be and our current apparatus is taking shape in attitudes of open and constant *control*' (2007, pp. 345–6), where power operates through technologies of tracking and connection which eliminate the need for brick and mortar enclosure. For Tippet, Deleuze's 'control' describes the post-panoptic surveillance/labour strategy of Facebook; for Nicholls, the modulation effected by wearable health motivation technologies reproduces the control of a system "in crisis"; and for Taffel, Deleuzian control is exemplified in the ruthless tracking efficiency of an Amazon warehouse. While Deleuze's 'Postscript' seems increasingly apt to describe the evolution of digital culture, it can also function incidentally to endorse the claims to "weightlessness" that these three chapters seek to refute. What they each show, with their individual deployments of Deleuze, is that a control society model of power is not one which precludes the implication of corporeality in its workings.

References

Deleuze, G. (1992). Postscript on the societies of control. *October, 59*(3), 3–7.

Deleuze, G. (2007). What is a dispositif? In D. Lapoujade (Ed.), *Two regimes of madness: Texts and interviews 1975–1995* (A. Hodges, & M. Taormina, Trans.). Los Angeles: Semiotext(e)

5

Corporate Geocorpographies: Surveillance and Social Media Expansion

Ryan Tippet

In August 2013, a collective of experts, local telecommunications services, and tech industry giants led by Facebook united under the banner of universal connectivity. Heralded by Mark Zuckerberg's white paper, 'Is Connectivity a Human Right?', the coalition both named and located online at 'Internet.org' announced its mission triumphantly: to bring affordable internet access to the two thirds of the world without it. Zuckerberg shows in the white paper, released to coincide with the launch of Internet.org, that although internet connectivity is spreading exponentially, the cost of overpriced mobile data on struggling infrastructure continues to exclude wide populations from participating in the 'global knowledge economy' (2013, p. 2). By making connectivity ten times as cheap, and phone applications ten times less data-intensive, Zuckerberg and his partners at Qualcomm, Ericsson, and Samsung, to name a few, aim to make a truly global internet a reality.

R. Tippet (✉)
Department of Media, Film and Communication, University of Otago, Dunedin, New Zealand

© The Author(s) 2016
H. Randell-Moon, R. Tippet (eds.), *Security, Race, Biopower*,
DOI 10.1057/978-1-137-55408-6_5

The popular critique of Internet.org notes pointedly that no one stands to benefit more from such a reality than the corporate partners that make up the collective (see Imtiaz 2014; Best 2014; Lopez 2014). Internet.org can be seen to fit within Facebook's long-term strategy to 'own' its users' first contact with the internet (Mims 2012). The name assumed by the collective exemplifies this strategy as 'Internet.org' becomes indissociable from 'the internet.' By ensuring that their company names are synonymous with the internet itself for emerging markets in Africa, Asia, and South America, Internet.org's corporate constituents aim to monopolise and manage a resource as they work simultaneously to make that resource indispensable. Further, Internet.org's full moniker, 'Internet.org by Facebook,' indicates the extent to which the 'non-profit' can be conceived as a branch of Facebook aiming to colonise untapped overseas markets. In fact, the issues which most undermine Internet.org's ostensible charity are the conflicting interests of its leadership. The effort in this chapter to foreground problems of ownership, following Mark Andrejevic, 'counters the determinism of those who insist on the *inherently* empowering character of interactive networks and the revolutionary *telos* of the digital era' (2007, p. 299). I situate the operations of Internet.org as an economic strategy in a charitable disguise, compromised by its corporate leadership and contradictory perspectives on property in the 'knowledge economy.' In short, and contrary to its benevolent rhetoric, Internet.org produces spaces of exploitation and surveillance in its efforts to establish an alluring market presence in underdeveloped parts of the world.

This chapter draws from surveillance studies conceptualisations of social media in concert with Joseph Pugliese's concept of 'geocorpographies' to formulate a critique of Internet.org as *corporate geocorpography*. I characterise corporate geocorpography, in parallel with Pugliese's (2007, p. 1) definition, as the *seductive and inclusive enmeshment of the flesh and blood of the body within the economic geography of race, technology, and imperialism*. Extending Pugliese's concept in this way helps to describe how Facebook works, through Internet.org, to produce a corporatised space which captures the value produced by its inhabitants through a surveillant business model of free labour. Thus, Internet.org's primary function is one of long-term economic gain for its partners, capturing new populations in what Andrejevic terms a 'digital enclosure' (2007). The first section of the chapter establishes the surveillance theory approach to Internet.org before the second examines the notion of digital enclosure in relation to the collective. The final section mobilises the concept of corporate geocorpography in relation to Internet.org, focusing on its public relations and marketing material to show how its ostensible charity is racially and colonially predicated upon the very bodies of its users and is a primarily economic endeavour thinly veiled by altruistic rhetoric.

The key product of Internet.org's efforts so far, and the central focus of critique in this chapter, is the 'Internet.org' Android app released in Zambia, Tanzania, and Kenya in 2014, with 'aggressive' aims to launch in 100 countries by the end of 2015 (Murphy Kelly 2015). Thanks to agreements with local telecommunications companies and service providers, the app gives users free access to a shortlist of services such as weather information, news sites, Wikipedia, and of course, Facebook itself.[1] The app has drawn the ire of net neutrality activists because it gives Internet.org tremendous power over how emerging markets use the internet, for what purposes, and with whose services. This is contrary to the central net neutrality principle that all data online should be treated equally by access providers (Wu 2003). The inclusion of Facebook's mobile app within the Internet.org app collection is a given, for instance, while its social media competitor Twitter is notably absent, privileging access for Internet.org users to one service, and denying it to another.

Surveillance Critiques of Social Media

It is taken for granted that contemporary Western societies are 'surveillance societies'—a loaded term which is not intended to convey a tone of 'sinister conspiracy' but simply to describe a feature of social life today (Lyon 2007, pp. 7–8; Ball et al. 2006). The computerisation of our work lives, social lives, public and private lives, lives as consumers, producers, and citizens facilitates an immanent and automatic form of recording and observation, which pervades with more depth and breadth than even George Orwell envisaged. Surveillance is an organising principle in our everyday lives, one which (quite literally) exposes us to asymmetries of power with the same readiness that it affords us the conveniences of social media, online banking, e-commerce, security, and an endless list of other luxuries. Surveillance's many uses, moralisations, and manifestations make it a complicated process to pin down, and it is increasingly inadequate to subsume all surveillance under a single concept, such as the enduring panopticon, the emblem of Michel Foucault's disciplinary societies which has long been the dominant model in critical surveillance research (Haggerty 2006). Certainly, it has become deeply outmoded within surveillance studies to approach the phenomenon of surveillance as a strictly negative one, despite this being the tradition since *Nineteen Eighty-Four* (Orwell [1949] 2003) and Foucault's ([1977] 1995) *Discipline and Punish* (Haggerty and Ericson 2000). Surveillance is configured as productive and necessary for security and a host of vital services in the digital age, but it remains essential that surveillance regimes be subject to critique. This section outlines concepts from the field of surveillance studies, which are central to the understanding of social media as a surveillant economic apparatus. I use these concepts to configure Internet. org as a post-panoptic surveillance technology, performing a function of economic dataveillance that captures the free labour of web users and operates within an assemblage of surveillance tools, strategies, and beliefs.

Kevin Haggerty and Richard Ericson (2000) have urged for new theorisations of surveillance, themselves drawing from Gilles Deleuze and Felix Guattari's concept of *assemblage*. The 'surveillant assemblage' accounts much more comprehensively for the complex and varied nature of surveillance today, as well as its tendencies towards 'function creep' (the use of innocuous tools in less innocuous surveillance settings) and the convergence of information across formats, institutions, and nations. Deleuze and Guattari coined the 'assemblage' in *A Thousand Plateaus* (1987), a concept summarised by Paul Patton as 'a multiplicity of heterogeneous objects, whose unity comes solely from the fact that these items function together, that they "work" together as a functional entity' (1994, p. 158). For surveillance, this means two things: that the diverse technologies of 'dataveillance' (Clarke 1988) we interact with every day also interact with one another to produce unexpected results; and that these technologies, along with the ideologies, attitudes, and perceived social transactions that support them, even when they operate disparately, contribute collectively to an overall state of hypervisibility in surveillance societies. The surveillant assemblage 'operates by abstracting human bodies from their territorial settings and separating them into a series of discrete flows. These flows are then reassembled into distinct "data-doubles" which can be scrutinised and targeted for intervention' (Haggerty and Ericson 2000, p. 606). The data-double is more than a digital representation of the human subject, it is 'increasingly the object towards which governmental and marketing practices are directed', and thus enables institutions to 'make discriminations among populations' (Haggerty and Ericson 2000, p. 613). It is useful to approach Internet.org as a technological and ideological node in the surveillant assemblage because the project is clearly more than just a surveillance mechanism. Although it is also a long-term strategy to elevate the importance of the internet in new markets, a collaboration of very different groups from a variety of sectors, and no doubt a genuinely useful resource in the lives of many users, Internet.org works to extend the reach of the surveillance society. The surveillant assemblage also underlines the importance of a particular political-economic and cultural context to Internet.org, namely, that of free market globalism and technological imperialism.

This surveillance theory approach to social media can be refined further with reference to post-panopticism. Surveillance in Deleuze's 'Societies of Control' (1992) constitutes a salient counterpoint to Foucault's disciplinary panopticon ([1977] 1995), bringing surveillance theory in line with digital technology and late-capitalist contexts—something Foucault is often criticised for having failed to do (see Poster 1990; Gandy Jr. 1993). Surveillance in the societies of control does not depend on locking its subjects within an unverifiable gaze. Instead, always-on tracking technologies and the neoliberal blurring of work/leisure boundaries dismantle the institutional interiority that marked and compartmentalised disciplinary societies, and automated dataveillance blankets the intervening space. The surveillant assemblage and Andrejevic's concept of the digital enclosure also challenge the universality of the panopticon metaphor and accord neatly with this notion of what Deleuze would term 'modulatory' control surveillance.

Foucault himself recognised the 'transience' of his model of power and saw on the horizon of a governmental analysis 'the image, idea or theme-program of a society in which there is an optimisation of systems of difference, in which the field is left open to fluctuating processes, in which minority individuals and practices are tolerated' (2008, pp. 259–60). It is primarily in this manner that what we may term 'economic' surveillance (versus 'political' surveillance [Fuchs 2011, p. 123]) operates—not through a physically coercive disciplinary normalisation aimed at diminishing difference in the production of docility, but through a 'fluctuating' process that savours difference, capitalises upon it, and channels it within a personalised and seductive marketing strategy. The control strategy of governance therefore depends upon finely-grained surveillance of private interests, desires, and inhibitions. In short, economic surveillance arguably does not aim to normalise—beyond the level of obligating consumption—but to identify the details that separate us from one another in order to sell acutely-targeted commodities. Andrejevic characterises this economy as one of 'mass customisation' based upon 'ever more precise forms of consumer surveillance that allow for individualised marketing and production' (2004, p. 2). In Siva Vaidhyanathan's words, 'These companies are devoted to tracking our eccentricities because they understand that the ways we set ourselves apart from the mass are the things about which we are most passionate' (Vaidhyanathan 2011, p. 112). Internet.org reflects a corporate inclination towards instituting such a consumption and customisation-based governmentality in emerging markets.

The final element of this surveillance studies schema comes from critiques of the 'free labour' of internet usage. Tiziana Terranova's (2000) paper on the subject, drawing on and extending the work of Maurizio Lazzarato (1996), is regularly cited to show how surveillance of personal data and web 2.0 participation renders 'interactivity' into free labour, generating surplus value for the likes of Facebook and Google. 'Immaterial labour', according to Lazzarato, includes the production of the 'cultural content' of commodities—things like 'defining and fixing cultural and artistic standards, fashions, tastes, consumer norms, and … public opinion' (1996, p. 132), which are often scarcely recognisable as 'work'. Terranova builds on this, noting that the internet in particular facilitates immaterial labour which does not appear as labour at all, but as a hobby or social practice. Her conclusion, that 'The pervasiveness of such production questions the legitimacy of a fixed distinction between production and consumption, labour and culture' (2000, p. 35), has only grown more relevant in the subsequent sixteen years, as the emergence of titanic social media web corporations with immense annual revenues demonstrates. When the interactions of Facebook's 1.39 billion monthly active users generate revenue totalling US$12.47b in one financial year (Facebook 2015), there can be little question that the platform's stream-lined interface and egalitarian rhetoric masks a subtle process of production and commodification that makes labourers of its participants. Terranova describes in her paper a form of dedifferentiation, which is echoed by Andrejevic who writes, 'The economic potential of digital dedifferentiation isn't based solely on allowing employees to continue working outside of the work space proper, but also to render non-work-related activities economically productive to the extent that they can be monitored' (2004, p. 35).

It has to be noted that Internet.org is by no means *primarily* a sur-veillant apparatus, but a technology for integrating new populations within the 'knowledge economy' (Zuckerberg 2013). However, there is no doubt that in the long-term, Internet.org does aim to expose these emerging markets to the dataveillance that already pervades in the rest of the world by enabling processes of 'bioinformationalisation' (see Chap. 1, Pugliese, this volume) and computerisation to take a hold of them. Even if it is a long-term strategy, its intended effects are already clear given the inclusion of Facebook on the Internet.org app and the privacy policy that governs the app (which redirects users to Facebook's own policy, doing little to dissuade anyone that Internet.org is best conceived as a branch of its founding parent company). For new users in Zambia, Kenya, Ghana, and many other nations, Internet.org heralds an internet for surveillance and commodifying free labour from its very beginning. If there is any doubt that Internet.org's is a primarily economic operation, consider the parallel project of one internet giant notably absent from the collective: Google's own 'Project Loon' aims to put balloons providing 4G coverage into low-Earth orbit, which—like Internet.org—will offer internet access in cooperation with participating mobile providers.[2] Both Google's and Facebook's efforts market themselves as socialistic and egalitarian tools of what Pierre Levy (1997) terms 'collective intelligence,' but they are also both initiated by corporations beholden to the demands of stockholders, and with vested interests in extending the internet's reach. Internet.org's corporate members and Google would each find some measure of benefit from the other's success, but both parties would also prefer to hold exclusive power over access to the internet. This is precisely what Andrejevic's concept of digital enclosure, discussed in the next section, demonstrates.

Internet.org as Digital Enclosure

Andrejevic (2004) shows that a process of 'digital enclosure' reproduces the enclosure of the agrarian commons described by Marx ([1867] 1976). By denying them access to the means of production, the enclosure of common land made workers dependent on capitalist enterprise for subsistence; they were "free" to become wage labourers. Digital enclosure, for Andrejevic, 'can correspondingly be understood as the process whereby activities and transactions formerly carried out beyond the monitoring capacity of the Internet are folded into its virtual space' (2004, p. 35). This means that computerisation and the growing dependence on 'cloud' services institutes a surveillant gateway between users and their personal information, their means of creativity and expression, and their social relations. Where a network technology aims to facilitate 'ubiquitous interactivity' (Andrejevic 2007, p. 296), it produces digital enclosure. Both agrarian and digital enclosures, then, 'compel entry into a specific space within which surveillance can take place' (Andrejevic 2004, p. 36)—first the panoptic factory, where the constant threat of surveillance effected an 'automatic' functioning of disciplinary power (Foucault [1977] 1995, p. 201), and then the networked and modulatory space of leisure, consumption, and communication produced by the internet. The two forms of enclosure differ in several key ways, as the land enclosure administers labour specifically whereas the digital enclosure encompasses the spheres *outside* of labour (and incites productivity within them). They also offer different incentives for entry: land enclosure involved the threat of physical violence, but digital enclosure, as the post-panoptic perspective stresses, trades on a softer form of violence—as desire, modulation, and seductive consumerism. That said, Andrejevic finds participation in the digital enclosure increasingly difficult to avoid or resist, noting that, 'taken to its limit, the logic of digital enclosure works toward the goal of foreclosing offline consumption options' (2004, p. 37). This ultimate vision of dependency on the internet for all consumption is especially concerning if power over the digital enclosure is centralised in the hands of corporate enterprise.

Andrejevic uses the spatial image of enclosure to counter the 'misleading' metaphor of networked services as a 'cloud' (2007, p. 297). The metaphor of the cloud evokes weightlessness, liberation, and constancy and accessibility of information services comparable to the oxygen we breathe. These connotations, Andrejevic claims, 'obscure the very concrete shifts in control over information associated with the recentralisation of information and communication resources envisioned by the architects of the internet "cloud"' (2007, p. 296). The notion of enclosure therefore reflects Andrejevic's emphasis on ownership as the central problem of commercial internet services. This focus on ownership also justifies the application of digital enclosure in critiquing Internet.org because 'within the digital enclosure those who control the resources—in this case, information-gathering technologies and databases—can lay claim to the value generated by those who enter "freely" into the enclosure' (2007, p. 314). Internet.org, in other words, institutes a barrier to the digital means of production, which enables Facebook to claim ownership over the value produced within its enclosure.

Andrejevic's concept of digital enclosure coheres well with the surveillance studies approach detailed above, as its emphasis on the extraction of value-producing data at the moment of interactivity aptly describes the processes of economic dataveillance and the surveillant assemblage abstraction of bodies from their 'data-doubles.' Although the image of 'enclosure' may seem to conflict with Deleuze's emphasis on porosity and openness in the societies of control, Andrejevic's point is that the technologies of modern surveillance operate beyond the pale of institutional interiors, so that their subjects are always locked within a surveillant gaze. For Andrejevic, entering digital enclosures means doing 'the work of being watched', a concept that mirrors Terranova's 'free labour' in describing the way activities not normally considered 'work' become commodified and exploited (Andrejevic 2004). The work of being watched is undertaken wherever "new media" aims to be interactive, for instance when reality television shows invite viewers to text in votes for their next democratically-elected pop idol. With this example, mass communication flows might claim to be democratic, allowing viewers to participate in its production and thus upending the one-sided relation of mass media. The real extent to which televised celebrity elections have liberated their audiences from the mostly one-way flow of mass media is questionable, but more importantly, Andrejevic finds that 'interactivity functions increasingly as a form of productive surveillance allowing for the commodification of the products generated by what I describe as the work of being watched' (2004, p. 2). The digital enclosure is an important stepping stone in the formulation of corporate geocorpographies because it emphasises the way a set of technologies associated with 'gaseousness' (Deleuze 1992, p. 4) and ethereality have real world manifestations and consequences for geographic space and corporeal bodies.

In the long term, Internet.org grants its corporate members tremendous power as gatekeepers of the digital enclosure being instituted in more regions of the world. Furthermore, by extending the digital enclosure through the charitable front of Internet.org, Facebook is widening its surveillant and exploitative business model to encompass emerging markets under the paradoxical banner of aid and altruism. Zuckerberg's belief in the power of the internet to solve resource and governance issues in poor nations may be genuine, but the lingering corporate spectre of Facebook stains the Internet.org mission and evinces an economic interest in these populations as new customers. Internet.org is from the beginning a form of digital enclosure, albeit loss-leading and nascent, and it is a myopic and naïve perspective which fails to see the project of a monolithic Western corporation to recruit its next billion members from impoverished nations for what it truly is: imperialistic co-option and globalist exploitation dressed up as charity.

Corporate Geocorpographies

Zuckerberg (in Chang and Frier 2015) has insisted that profit is not a priority for Facebook's expansion into other countries through what might be characterised as the Trojan Horse of Internet.org. However, the 'Focus on Efficiency' white paper by Facebook, Qualcomm, and Ericsson notes on the continued use of low-powered 'feature phones' over smartphones, that '"Facebook for Every Phone" became a profitable service despite the doubts regarding the monetisation prospects of its demographics' (Internet.org 2013a, p. 43). This means that although the move to make a stripped-down version of Facebook available for outdated devices owned by poorer demographics was seen to be a "loss-leader," Facebook managed to generate a profit from those demographics with the Facebook for Every Phone program. Exactly how this occurred is unclear (especially as Facebook for Every Phone's framework does not support advertising) and not elaborated upon, but the platform's profitability may be a broader reflection of the market penetration and mass data-collection enabled by Facebook's feature phone compatibility. At any rate, this contradicts Zuckerberg's repeated claim that the poor do not constitute a profitable demographic for Facebook, and it makes it difficult to deflect the argument that Internet.org aims to take advantage of poor and under-resourced international markets. Ericsson and Qualcomm, in that same paper, are even less sensitive to the critiques of Internet.org as a profit-seeking machine in references to the 'expanding market opportunity' it provides (Internet.org 2013a, p. 54).

Internet.org's economic expansionism is predicated on assumptions that internet technology will be unproblematically adopted with unanimously positive benefits in new regions. The 'revolutionary *telos*' (Andrejevic 2007, p. 299) of digital technology is never questioned, but ascribed a place of inherent progress and betterment in claims that 'we believe everyone deserves to be connected' (Zuckerberg 2013, p. 1), and 'a knowledge economy is different [from a modern industrial economy] and encourages worldwide prosperity' (p. 2). No space is left in this rhetoric to question the tacit assumptions of progress bound up in global free market expansion. As Andrejevic instructs, the very fact of corporate involvement in this 'revolution' 'suggests heads won't be rolling any time soon' (2004, p. 15). Zuckerberg refers frequently to the 'knowledge economy' as a property-less utopia, noting with oblivious irony, 'If you know something, that doesn't stop me from knowing it too' (2013, p. 2). The infinitely reproducible nature of information and digital commodities promises a more egalitarian world of collective intelligence, Zuckerberg writes: 'In fact, the more things we all know, the better ideas, products and services we can all offer and the better all of our lives will be' (2013, p. 2). This claim that the knowledge economy miraculously dispenses with the ills of property is problematic and contradictory for two reasons. First, Facebook's business strategy depends precisely on being able to take ownership of its users' personal information, and second, this ethos of freely reproducible—economically valueless—information is, not coincidentally, advantageous to Facebook as the master and gatekeeper of the Internet.org digital enclosure in the control society context of free labour. In the knowledge economy that Facebook aims to facilitate, information *is* a valueless commodity—except for those who are in a position to render it profitable. This celebratory conception of the knowledge economy promises collective benefits, but within the digital enclosure established by Internet.org, it instead redraws imperial and capitalistic relations of power, paralleling past colonial land enclosures with a seductive and inclusive rather than disciplinary and exclusive tact.

Internet.org produces its subjects discursively as racialised Others and an underutilised resource. Depictions of the imagined users of Internet.org services in promotional material lean heavily on stereotypical cultural tropes to mark their bodies as racially and culturally different, and the narration and images frequently imply a distinct "lack" in the way they live—a lack of the sophistication and self-sufficiency the internet has granted the West (see Internet.org 2013b, 2015a, b, c, d). Furthermore, this material positions Internet.org users as aspirational individuals who need only be given the tools to make their way in the free market. Voiceovers in advertisements describe how users become a valuable intellectual resource and can contribute to the collective intelligence of the global population—in other words, we all benefit from bringing the poor and disadvantaged online. Internet.org's techno-imperialist assumptions of miraculous prosperity (bound in seductive inclusivity) are articulated through a neoliberal logic of aspiration and responsibilisation that envisions each user as entrepreneurial *homo economicus* (Foucault 2008). The thirty-second advertisements hosted on Internet.org's YouTube channel are especially frank in this regard, each profiling a different person or pair of people from a country targeted by the 'Internet.org' app. The videos emphasise how their lives would be improved by access to the internet. For example, brothers Mehtar and Mostek, it is claimed, had to independently invent and design the windmill because they were out of the knowledge economy loop. The ridiculous notion that they would be so isolated from civilisation that they would need to 'figure it out for themselves, because they're not connected to the internet' (Internet. org 2015c), is rendered farcical in shots of one brother using a welder to attach pieces of their windmill. The tacit assumption is that the profiled individuals would be best served by access to the tools of participation in the knowledge economy, rather than by government or direct aid. Further, the narrative leans on racist generalisations and stereotypical images of the geopolitical South as underdeveloped and primitive. This is evident in the way that Internet.org more broadly produces the knowledge economy with a favourable inclination towards a Western epistemology in which technical, scientific, rational, and "expert" knowledge is sacrosanct and superior. Non-Western epistemologies of knowledge-generation are diminished in these texts by the implicit narrative that the technical knowledge employed, for example, by Mehtar and Mostek is the answer to all the developing world's problems.

The brothers' windmill is seen in one of the final shots of the advertisement helping to pump water out of a well for locals. This is Internet.org's answer to critics who would sooner see poorer nations provided with running water than internet access. The message of responsibilisation is clear: with access to the internet, people can 'figure out' running water for themselves. Because the target audience of these advertisements is not the people who will be relying on Internet.org, each advertisement highlights the benefits the rest of us take from widening the internet's digital enclosure: 'And with all [Mehtar and Mostek's] ingenuity and resourcefulness, think of what they might share with us' (Internet.org 2015c); and in 'Erika & Esmeralda', 'What else could they build? It could be anything. And that's why we need to connect them. Because the world needs their ideas and creativity' (Internet.org 2015a). Not only will involvement in the knowledge economy better the lives of each video's subject, but it will contribute to the collective intelligence of the whole network, the videos claim, each closing with the maxim, 'The more we connect, the better it gets.' As mentioned above, this ethos of universal connectivity happens also to coincide with the tenets of control societies, of digital enclosure, and of Facebook's mass customisation business model. The YouTube ads epitomise the rhetoric of a 'silicon revolution that painlessly eliminates the inequities attendant upon the concentration of control over wealth and productive resources by economic and political elites,' as Andrejevic describes it, noting that 'The key to this hypothetical revolution is not the redistribution of control over material resources, but their supposed irrelevance in the emerging information economy' (2007, p. 299).

In 'Is Connectivity a Human Right?', Zuckerberg describes the problem of 'aligning incentives', a euphemistic phrase which means, essentially, crafting a plan for free internet profitable enough that corporations, internet services, and mobile operators will be enticed to contribute to it, yet attractive enough that users will be seduced to participate. Facebook sees its own role as that of a facilitator—it puts the right companies and local governments in conversation with an aim to line up their interests for expanding internet access. All this is to say that the problem of connectivity as an essential human right is one which, to Internet. org, is both discursively produced and cynically resolved by global capital with Facebook at its helm. The question of whether or not the resources necessary to provide this human right should be provided by capitalist institutions acting as non-profit charities does not enter the discussion. Internet.org and its proponents may assert that no other possibility for instituting worldwide connectivity exists—that the infrastructure needs to be put in place, and the only people willing to fund it are those who can make it profitable in the near future; in other words, we cannot wait for the Zambian or Guatemalan government to provide the human right of connectivity. However, we must be critical of the way this rhetoric shuts out 'the possibility of alternative conceptions of how digital, interactive infrastructures could be developed and implemented' (Andrejevic 2007, p. 311). Similarly, the argument that Internet.org is purely altruistic because it is giving people something for free must be tempered by a recognition that the central function of enclosure is both to seduce and preclude alternative options.

Based on this analysis of Internet.org's racialised public relations output and the digital enclosure of the 'Internet.org' app, I contend that the collective is aimed at instituting a space of corporate geocorpography in emerging non-Western markets. For Pugliese, geocorpography refers to spaces which enable 'the violent enmeshment of the flesh and blood of the body within the geopolitics of race, war and empire' (2007, p. 1), for example the infamous Abu Ghraib prison in Baghdad, Iraq. The concept of geocorpography recognises that geopolitical flows produce spaces and bodies in ways distinctly inflected by their uneven power dynamics. The conceptual link between what are perceived traditionally as "hands-off" and intangible forms of dataveillance undertaken by social media and the violent bodily manifestations of geopolitics in geocorpographies comes from Andrejevic's digital enclosures and post-panoptic perspectives of desire and seductive non-disciplinary surveillance. At the intersection of economy and geocorpography, then, we find the dominant agents are not exclusivist states, but corporations seeking to include a global population among their customers; we find not the corporeal violence of torture or malnourishment, but the seductive incorporation of bodies within systems of free labour and digital enclosure; finally, we find not a search for productive and normative docility, as in Foucault's disciplines, as much as an imperialist *demarcation* and *incorporation* of racial, cultural, and technological difference.

It can be argued that in the context of globalised economy, powerful multinational corporations are increasingly aimed at accessing and co-opting states and polities. That is to say they have a growing interest in and power over how governments operate—for example, in lobbying for weak and exploitable labour laws. In Pugliese's conception, geocorpographic space is produced by states and the violence of nationhood, but I argue that corporations play a growing role in that production of geocorpography. So-called "sweatshop" factories governed by underpowered and inhumane labour laws could be described as geocorpographic spaces produced by state governance, but inflected or directed by corporate enterprise with vested interests in how such spaces are administered. In the case of Internet.org, there is also an effort on the part of corporate entities to influence state policies and governance—however the aim and method is less disciplinary than a sweatshop—and geared more towards the production of consumer-citizens than citizens as producers (although digital free labour blurs this distinction). Corporate geocorpographies produce spaces where the bodies of the populations in emerging markets are co-opted within economic regimes which render them productive for the capitalist owners of what might be termed the means of connectivity—

the digital enclosure. The fact that those gatekeepers of digital enclosure are predominantly Western corporations seeking to colonise new markets in "developing" nations lends corporate geocorpography an imperialistic propensity. Corporate geocorpographies reflect the context of both geopolitics and economic geography, marking subject bodies simultaneously as racialised Others to be delineated biopolitically and valuable resources to be co-opted indiscriminately.

Conclusion

This chapter has shown how Facebook's Internet.org project can be seen to operate in embodied and imperial dimensions which are absent from the "net neutrality" critique. To this end, Internet.org institutes a digital enclosure, configured more specifically as corporate geocorpography. The concept of corporate geocorpographies elucidates the way that not only states, but corporations can produce and delineate spaces within which individuals find themselves subject to a seductive enmeshment of global capital and corporeality. The digital enclosure of Internet.org implicates the very bodies of its users within an imperially-predicated relationship of neoliberal capitalist expansion, and Internet.org's marketing images and rhetoric frames those users both as an underdeveloped Other to be enlightened by Western epistemology and as an untapped resource within utopian conceptions of the knowledge economy, propelling the globe towards weightless labour. Using post-panoptic concepts from surveillance studies, namely, the surveillant assemblage, surveillance in societies of control, and immaterial labour as 'the work of being watched,' this chapter has shown how surveillance works for Facebook to enable the commodification and capture of "free labour." To the extent that Internet.org can be thought of as a branch of Facebook, then the ostensibly non-profit organisation is one node in the surveillant assemblage which creeps towards universality under the benevolent banner of worldwide connectivity.

The question posed by Zuckerberg in 2013—is connectivity a human right?—is not the question that needs to be asked, and not the one this chapter has aimed to answer. A more important question, for which there is no space in Internet.org's altruistic rhetoric, is this: *if* connectivity is a human right, should access to it be granted by a corporation with a surveillant business model? The promise of Internet.org, in its celebratory techno-determinism, is that of mass customisation as a form of democratic power-sharing. However, 'Viewed from a slightly different angle … the promise of power sharing reveals itself as a ruse of economic rationalisation and the promised form of participation comes to look a lot like work' (Andrejevic 2004, p. 6). This is surveillance's seductive character. The promise of Internet.org, of inclusion, revolution, and the redistribution of power, and of egalitarian participation in the boundlessly fruitful knowledge economy, comes with a tacit and unspoken entry fee: privacy, cultural independence, an alternative framework for access, and even— when the concept of corporate geocorpography is taken its furthest—a degree of corporeal freedom.

Notes

1. In response to net neutrality controversy over the app in India, Internet.org has recently widened the scope of the app to the Internet. org platform, renamed to 'Free Basics' in September 2015, where companies can apply to be added to the list of free services on offer. They are, however, still subject to the limitations imposed by the collective, and the app still institutes a divide between 'zero-rated' free services and paid ones which are ineligible for the platform.
2. Another project, code-named 'Titan,' aims to do the same with solar-powered drones that were acquired when Google outbid Facebook in 2014 for Titan Aerospace.

References

Andrejevic, M. (2004). *Reality TV: The work of being watched*. Lanham: Rowman & Littlefield Publishers.
Andrejevic, M. (2007). Surveillance in the digital enclosure. *The Communication Review, 10*(4), 295–317.

Ball, K., Lyon, D., Murakami Wood, D., Norris, C., & Raab, C. (2006). *A report on the Surveillance Society for the Information Commission by the Surveillance Studies Network*, D. Murakami Wood (Ed.). http://news.bbc.co.uk/2/shared/bsp/hi/pdfs/02_11_06_surveillance.pdf. Accessed 29 May 2015.

Best, M. (2014). The internet that Facebook built. *Communications of the ACM, 57*(12), 21–23.

Chang, E., & Frier, S. (2015). Mark Zuckerberg Q&A: The full interview on connecting the world. *Bloomberg Business*, 19 February. http://www.bloomberg.com/news/articles/2015-02-19/mark-zuckerberg-q-a-the-full-interview-on-connecting-the-world. Accessed 28 Apr 2015.

Clarke, R. (1988). Information technology and dataveillance. *Communications of the ACM, 31*(5), 498–512.

Deleuze, G. (1992). Postscript on the societies of control. *October, 59*(3), 3–7.

Deleuze, G., & Guattari, F. (1987). *A thousand plateaus: Capitalism and schizophrenia* (B. Massumi, Trans.). Minneapolis: University of Minnesota Press.

Facebook. (2015). Facebook reports fourth quarter and full year 2014 results. *Facebook*, 28 January. http://investor.fb.com/releasedetail.cfm?ReleaseID=893395. Accessed 28 April 2015.

Foucault, M. (1977/1995). *Discipline and punish: The birth of the prison* (A. Sheridan, Trans.). New York: Random House.

Foucault, M. (2008). *The birth of biopolitics: Lectures at the Collége de France, 1978–79*. G. Burchell, Trans. & M. Senellart (Eds.). New York: Palgrave Macmillan.

Fuchs, C. (2011). How can surveillance be defined? *MATRIZes, 5*(1), 109–133.

Gandy, O. H., Jr. (1993). *The panoptic sort: A political economy of personal information*. Boulder: Westview Press.

Haggerty, K. D. (2006). Tear down the walls: On demolishing the panopticon. In D. Lyon (Ed.), *Theorizing surveillance: The panopticon and beyond*. Devon: Willan Publishing.

Haggerty, K. D., & Ericson, R. V. (2000). The surveillant assemblage. *British Journal of Sociology, 51*(4), 605–622.

Imtiaz, A. (2014). Nothing altruistic about Facebook's initiative to spread the Internet. *US Finance Post*, 6 January. http://usfinancepost.com/nothing-altruistic-about-facebooks-initiative-to-spread-the-internet-11862.html. Accessed 27 May 2015.

Internet.org. (2013a). A focus on efficiency: A whitepaper from Facebook, Ericsson and Qualcomm. *Facebook*, white paper, 16 September. https://fbcdn-dragon-a.akamaihd.net/hphotos-ak-prn1/851575_520797877991079_393255490_n.pdf. Accessed 28 Apr 2015.

Internet.org. (2013b). Every one of us. #ConnectTheWorld. *Youtube*, online video, 20 August. https://www.youtube.com/watch?v=NdWaZkvAJfM. Accessed 19 May 2015.

Internet.org. (2015a). Erika & Esmeralda. *Youtube*, online video, 25 February. https://www.youtube.com/watch?v=9ffWa6cOb7M. Accessed 19 May 2015.

Internet.org. (2015b). Lian. *Youtube*, online video, 25 February. https://www.youtube.com/watch?v=JdUovve48No. Accessed 19 May 2015.

Internet.org. (2015c). Mehtar & Mostek. *Youtube*, online video, 25 February. https://www.youtube.com/watch?v=zdRvlZY5F_c. Accessed 19 May 2015.

Internet.org. (2015d). Neesha. *Youtube*, online video, 25 February. https://www.youtube.com/watch?v=0F2M0XQp0-E. Accessed 19 May 2015.

Lazzarato, M. (1996). Immaterial labor. In P. Virno & M. Hardt (Eds.), *Radical thought in Italy: A potential politics*. Minneapolis: University of Minnesota Press.

Levy, P. (1997). *Collective intelligence: Mankind's emerging world in cyberspace*. New York: Plenum Press.

Lopez, A. (2014). The drones of Facebook (and the NSA). *Counterpunch*, 3 April. http://www.counterpunch.org/2014/04/03/the-drones-of-facebook-and-the-nsa/. Accessed 29 May 2015.

Lyon, D. (2007). *Surveillance studies: An overview*. Cambridge: Polity Press.

Marx, K. (1867/1976). *Capital: A critique of political economy, volume one* (B. Fowkes, Trans.). London: Penguin Books.

Mims, C. (2012). Facebook's plan to find its next billion users: Convince them the internet and Facebook are the same. *QUARTZ*, 24 September. http://qz.com/5180/facebooks-plan-to-find-its-next-billion-users-convince-them-the-internet-and-facebook-are-the-same/. Accessed 29 May 2015.

Murphy Kelly, S. (2015). Facebook wants to bring free web access to 100 countries by end of year. *Mashable*, 5 March. http://mashable.com/2015/03/04/facebook-web-access-internet-org/. Accessed 29 May 2015.

Orwell, G. (1949/2003). *Nineteen eighty-four*. London: Penguin Books.

Patton, P. (1994). MetamorphoLogic: Bodies and powers in *A thousand plateaus*. *Journal of the British Society for Phenomenology, 25*(2), 157–169.

Poster, M. (1990). *The mode of information: Poststructuralism and social context*. Cambridge: Polity Press.

Pugliese, J. (2007). Geocorpographies of torture. *Critical Race and Whiteness Studies, 3*(1). http://www.acrawsa.org.au/files/ejournalfiles/65JosephPugliese.pdf. Accessed 25 Aug 2015.

Terranova, T. (2000). Free labor: Producing culture for the digital economy. *Social Text, 18*(2), 33–58.

Vaidhyanathan, S. (2011). *The Googlization of everything: (And why we should worry)*. Berkeley: University of California Press.

Wu, T. (2003). Network neutrality, broadband discrimination. *Journal on Telecommunications & High Technology Law, 2*(1), 141–175.

Zuckerberg, M. (2013). Is connectivity a human right? *Facebook*, white paper, 20 August. https://fbcdn-dragon-a.akamaihd.net/hphotos-ak-xpa1/t39.2365-6/12057105_1001874746531417_622371037_n.pdf. Accessed 29 May 2015.

6

Everyday Modulation: Dataism, Health Apps, and the Production of Self-Knowledge

Brett Nicholls

Introduction: Dataism

The clear message of the health app industry is that wearable technology offers the optimum conditions for self-empowerment, self-management, and self-improvement. The celebratory pitch of this message is built upon new developments in mobile technology and experiments in new possibilities for the relationship between bodies and data. As Melanie Swan argues, in this relationship, the body 'becomes a more knowable, calculable, and administrable object', with an 'increasingly intimate relationship with data as it mediates the experience of reality' (2013, p. 85). Advocates for the benefits of intimacy with data, the California-based Quantified Self Labs contend that wearable technologies promise to open up new horizons for 'self knowledge through numbers' (Quantified Self 2015). This knowledge can be produced, as the Labs would have it, through the

B. Nicholls (✉)
Department of Media, Film and Communication, University of Otago,
Dunedin, New Zealand

H. Randell-Moon, R. Tippet (eds.), *Security, Race, Biopower,*
DOI 10.1057/978-1-137-55408-6_6

101

use of data-generating sensors both worn on the body and placed in as many locations as possible throughout our social spaces. And while the organisation is quick to point out that strict controls need to be in place to limit who has access to this data, their fundamental claim, along with the health app industry, is that a somatechnic reconfiguration of bodies, technology, and data is both desirable and socially necessary.

Joseph Pugliese and Susan Stryker write that the concept of somatechnics 'troubles and blurs the boundary between embodied subject and technologized object' (2009, p. 1) by refusing the common sense dichotomy of the subject/object, the self, and the world. It refers to the notion that the human body is not 'naturally occurring' as much as it is the 'tangible outcome of historically and culturally specific techniques and modes of embodiment processes' from which it cannot be dissociated (p. 2). A somatechnic perspective seeks to identify and describe the ways the human body is always already *produced* by the cultural and technical flows that engage it; in short, 'we have never existed except in relation to the *techné* of symbolic manipulation, divisions of labour, means of shelter, and sustenance, and so forth' (p. 2). Wearable technologies and health motivation apps are a particularly visible technology of somatechnic embodiment, which physically and discursively produce users within the specific cultural contexts of health and fitness, productivity, data-reliance, class, and culinary modes. I will consider in this chapter how the ideology of 'self knowledge through numbers' functions somatechnically to reconfigure everyday embodied practices as simple as walking and eating.

It is important to note that the claim by advocates of the quantified self, that knowledge of life and the body is improved through data, is not grounded in social and political analysis. Instead, the claim is based upon a version of what José van Dijck calls an 'ideology of *dataism*' (2014). This is an ideology, she maintains, driven by entrepreneurial desire as well as the mechanisms of state surveillance. Dataism involves the assumption that data consists of '*imprints* or *symptoms* of people's actual behavior or moods' (2014, p. 198; original emphasis). She explains,

> datafication has grown to become an accepted new paradigm for understanding sociality and social behavior. With the advent of Web 2.0 and its

proliferating social network sites, many aspects of social life were coded that had never been quantified before—friendships, interests, casual conversations, information searches, expressions of tastes, emotional responses, and so on. (p. 198)

Health motivation apps open up this paradigm to a more personalised relationship between selves, technology, and data. Users can carry self-produced data around in their pockets or bags, and this data provides the basis for reconfiguring everyday practices such as walking, eating, and sleeping. Crucially, van Dijck goes on to argue that the 'widespread *belief* in the objective quantification and potential tracking of all kinds of human behaviour and sociality … is rooted in problematic ontological and epistemological grounds' (p. 198). Dataism also involves, she maintains, a dubious '*trust* in the (institutional) agents that collect, interpret, and share (meta)data' (p. 198). The software and algorithms that analyse data are, for the most part, hidden from the user. The ideology of dataism is a necessary support, then, for reconfiguring everyday practices. The promise of 'self knowledge through numbers' thus begs several interrelated questions: how do wearable technologies situate selves in relation to data? How does the opening of "new horizons" connect with and reconfigure everyday practices? How do wearable technologies support or challenge existing regimes of power? In this chapter I will consider these questions by focusing on health motivation apps, such as Fitbit and Jawbone's UP24.

The promise of self-improvement through health motivation apps is big business. According to the *Research2Guidance*'s 2014 'mHealth App Developer Economics' report, there are more than 100,000 health apps on the market. The report states, 'market revenue reached USD 2.4bn in 2013 and is projected to grow to USD 26bn by the end of 2017' (2014, p. 7). The industry promotes itself as innovative, where slick technical design connects with personal desires to achieve fitness and other lifestyle goals. For instance, Fitbit claims to 'empower and inspire you to live a healthier, more active life'. Fitbit's aim is to design 'products and experiences that fit seamlessly into your life so you can achieve your health and fitness goals, whatever they may be' (Fitbit 2015). Azumio directly appeals to the virtues of quantification. The company slogan is 'Quantify your day-to-day'

(Azumio 2015a), and their flagship smart phone health app, branded Argus, aims to 'improve your wellness' (Azumio 2015b). The company's stated mandate is 'to inspire healthy habits through understanding and management. We want to make the personalization of healthcare available for everyone' (Azumio 2015b). And Jawbone employs a challenging slogan: 'There's a better version of you out there. Get UP® and find it' (Jawbone 2015a). 'Jawbone's UP® system', the company claims, 'helps people live better by providing personalized insight into how they sleep, move and eat. Its open Application Program Interface (API)—the UP Platform—includes an ecosystem of apps and services that integrate with UP to offer new, customized experiences' (Jawbone 2015b). As can be seen from these examples, the health app industry promotes an imperative of self-improvement facilitated by app-generated data.

What is at issue with the message of the health app industry is how data, as ontology, functions in everyday contexts. Health motivation apps require users to understand, as per an ideology of dataism, numerical information as an objective and precise index of bodily activity. This understanding marks a shift in the everyday role of language and other resources—including institutions such as media, school, family, and so on—in the construction of self-identity. As opposed to the ambiguity and judgement associated with the descriptive use of language ("that run was tough, I ran a long way today"), numbers are understood as an index of bodily activity ("I ran 10kms today, that's 27 kms so far this week"). In this example, the bodily experience of running is articulated by an app as a specific quantity. Running apps and step counters do not simply measure distance and count steps. Quantification facilitates two interrelated processes. First, the app tracks running across time, and, second, it also links the user to calculable formulae such as: the quantity of calories burnt is directly proportional to distance, speed, and body mass (Hall et al. 2004). Likewise, counting calories makes food intake meaningful in two interrelated ways. First, as quantity, food becomes understood as potential energy, and the body becomes a mass for expending energy. On the second level, calories are connected up and balanced within an input/output equation in order to maintain mass (food intake = energy output), or else reduce mass with an unbalanced equation (food intake < energy output).

The Health Motivation App

I am considering numbers here as causally determined signifiers that link the self to calculable formula because this is precisely how an ideology of dataism functions. As I stated, my aim is to consider how dataism works in relation to the reconfiguration of everyday practices around new somatechnic arrangements. The brief examples above show how selves can be inculcated in calculable formulae that shift perceptions of the body. But before considering the question of reconfiguration more fully, it will be useful to make some observations on how health motivation apps work. Health motivation apps can be described as sociotechnical measurement machines. Many consist of a sensor attached to the arm or shoe, which connects to a smart phone via Bluetooth. The corresponding smart phone app itself generates graphs and provides other forms of information on the basis of the data generated by the sensor and other inputs from the user.

The more comprehensive health motivation apps work in seven basic ways. I will focus on the Jawbone UP system app. First, users are required to input data about themselves: age, weight, and gender. These are rigid and abstract categories. Through this input, the user's body is effectively deracialised and situated via the conventional gender binary. The app reproduces what Joseph Pugliese has called the biopolitical effects of an 'infrastructural calibration of whiteness' (2010, p. 64). Second, users are asked to set goals: desired number of steps per day and/or weight loss aims. Third, the app then tracks specific aspects of daily life—hours slept, steps/distance, calorie intake/nutritional details, and sometimes mood. Fourth, these aspects are mapped and patterns of behaviour are charted across time. For instance, users can see how many steps they take on each day of the week; they can see if the number of steps is consistent, or if there are some days in which step activity is lower than others. Another example might be the user noticing that they eat more salt on Wednesdays in comparison to the rest of the week, or they sleep less on Fridays. Fifth, the apps evaluate these patterns in terms of statistical norms. Individual user data is situated in relation to other users and other institutionally-produced health standards. Users can discover, for instance, that in their age group they are ranked in the top 25 per cent for the number of steps taken per day, or they might get feedback that their daily salt intake is well above levels recommended by government-funded food science and health institutions. Sixth, the apps generate recommendations for subjects to modulate their practices. On the basis of their data, a user may decide to try to increase their number of steps per day, or reduce salt levels in their diet. They may notice that Friday's lack of sleep correlates with their after work alcohol intake, and try to drink more moderately. Seventh, the apps allow individual data to be viewed by and compared with others in competitive social networks.

What is crucial about the health motivation app is that two levels of data are produced. On the first level, the sensor/app arrangement generates a data set for individual use. Individual users can track how many steps they have taken over a period of time, hours slept, calories consumed, etc. They can then modify specific practices on the basis of this data: eat more healthily, walk more, sleep more, and so on. Data can also be tracked across time as an index of progress or decline. Users can set goals and clearly map their progress. As one typical user explains, the wearable sensor 'has made me aware of my actual progress throughout the day and in doing so, encouraged me to go to the gym ~3× a week so I actually hit my self-assigned step goal' (Amazon 2013). Moreover, individual datasets can be shared on social networks with competition as a motivator. Another user asserts, 'I like knowing my friends can see my steps because it provides greater motivation to be awesome' (Gray 2014).

On the second level, the individual's phone becomes one sensor among many others, and functions as an element in a massive data ecosystem that can be mined to produce knowledge and information about populations as a whole. Concerns about privacy are rightly raised with respect to this level of metadata (Andrejevic 2009), and there is a growing body of work engaging with implementing wearable technologies in health care systems (Lupton 2014; Mandl et al. 2015). The wearable technology industry is in its infancy, and no one knows what sociotechnical configurations will emerge in the future. What is clear is the assumption that social problems, defined in this context by the discourses of public health, can be solved through processes of datafication that connect with the vicissitudes of everyday life. My contention is that wearable biometric devices are new capillary technologies which directly connect the everyday, mundane practices of individuals to broader systems of social control.

Governmentality and Control Societies

The process of reconfiguring everyday practices of bodies, technology, and data is best understood in terms of Deleuze's prescient work on societies of control (1992). The question of control is consonant with the Foucauldian concept of governmentality (2009), which has been central to recent work dealing with the discourses of health. In one such critique of health promotion within a "neoliberal" context, Nike Ayo points out:

> the concept of governmentality provides a useful tool for demonstrating how health promotion works, not by making social and structural changes which impede upon the health and wellbeing of the population, but rather, by inciting the desire within autonomous individuals to choose to follow the imperatives set out by health promoting agencies, and thus, take on the responsibility of changing their own behaviours accordingly. (2012, p. 100)

Following Ayo, the health app can be seen to link the imperatives of health-promoting agencies to the sphere of everyday life. This technological linkage provides the basis for the neoliberal discourse of self-responsibility to take effect. Through its institutionally-produced recommendations for eating, moving, and so on, the health app lays the groundwork for the conduct of users. The health app deploys the actuarial logic that Foucault understands as security (Gordon 1991, p. 45). Wearable technologies provide security within one's body against the threat of heart disease, cancer, etc. At the same time, individual users categorise themselves on the basis of age, weight, and gender and choose goals in relation to the app's recommendations. "Choice" here thus takes the form of a wager. The degree to which users align with the technology is directly proportional to the desired level of security. The curious effect is that individuals are empowered to make choices that have already been mapped out for them. In Foucault's terms, this is precisely how the mechanism of what he calls 'biopower' works as an 'indispensable element in the development of capitalism'. Biopower is a method of 'power capable of optimizing forces, aptitudes, and life in general without at the same time making them more difficult to govern' (1990, pp. 140–141).

I want to extend Ayo's Foucauldian insights to Deleuze's (1992) work on control societies. Deleuze's ideas on control resonate with Foucault's account of governmentality and allow us to see how dataism reconfigures everyday practices. If we follow Deleuze's thinking, control somatechnically manifests in the intermeshing of wearable technologies and subjects through a series of directing, calculating, and ideological mechanisms of power. Everyday life is (re)configured through the governing of conditions of possibility, and the future potential of users is linked by apps to previous practices that have been recorded and subjected to abstract and calculated accounts of the healthy and productive body. Deleuze's concepts of modulation and networks explain how this control process works and why dataism is crucial for it.

For Deleuze, societies of control consist of volatile and dispersed networks, as opposed to the disciplinary societies outlined by Foucault (1977), which were structured around enclosed institutions such as the factory, hospital, school, prison, and barracks. Crisis is a central feature of control societies: in control societies corporate and public institutions are constantly restructured and fine-tuned. No existing combination or organisation can be considered permanent. As Deleuze puts it, 'in societies of control one is never finished with anything—the corporation, the educational system, the armed services being metastable states coexisting in one and the same modulation, like a universal system of deformation' (1992, p. 5). In the case of wearable health motivation technologies, this modulation is precisely what we find. Subjects modulate behaviour on the basis of data drawn from different aspects of everyday life and information from health corporations.

Deleuze tells us a control mechanism works by 'giving the position of any element within an open environment at any given instant' (1992, p. 7). Such mechanisms don't work to determine outcomes in terms of a predetermined norm, they work to measure states so that practices can be modulated to enhance efficiency and maximise outputs. The rationale for self-knowledge through numbers assumes data—as much data as possible—about the system will lead to more productive and happier lives. The user/app interface is a cybernetic feedback loop in which the body is reduced to a nodal point in a system of communication connected to larger networks. What matters is the capacity of selves to connect to forming and deforming institutions and systems of information. If health motivation apps communicate one clear message, it is that modulation and alteration are vital aspects of contemporary life.

Modulations

Let us turn now to consider health motivation apps in relation to modulation. Processes of modulation involve, on the one hand, limitless possibilities for combining elements in and across systems to maximise efficiency and, on the other hand, an ever increasing network of data-gathering sensors upon which to construct predictive models. Undeviating and fixed states are ideological and technological anathema in societies of control. At the most basic level, the user engages with a health app that manifests the central features of control societies. Apps are constantly in the process of updates designed to fix software bugs, improve usability, and introduce new features. The use of the app must be configured around this ever-changing technology. Moreover, the first issue that confronts users is how to get the app working properly. The integration of a sociotechnical measurement machine into everyday life is not without difficulties. It requires subjects to wear a band on the arm, or device on the shoe, which needs to be regularly charged for the Bluetooth capability to function. This band has to be manually switched into sleep mode, in the case of the Jawbone UP system, so that the device can monitor sleep patterns. The technology also requires users to modulate the gait of their walking: in order for steps to be counted accurately the arm has to swing. If users walk with their hands in their pockets, or they push a shopping trolley, the device may not count steps. Recording data on diet requires users to scan bar codes from packets of food, or enter food information directly. In the case of the Jawbone system, mood indicators also need to be manually selected from a continuum of six possibilities from feeling 'exhausted' to feeling 'fantastic' at various points throughout the day.

In order to generate accurate, longitudinal data, the user must actively reconfigure their practices in relation to the technology. This involves subtle modulations of the body and everyday practices. In this process, everyday life is ordered according to a new classificatory schema (Desrosières 1998). Particular aspects of everyday life—steps, sleep, diet, and mood—become points of greater focus, as opposed to other aspects of everyday life. In fact, other aspects of everyday life—such as work, sex, listening to music, gardening, and so on—are subordinated within a hierarchical relationship to the key aspects of the health motivation app. Improvement in the key identified areas will lead to improved performance in the other aspects of users' lives. Herein lies the promise of the health motivation app: increased productivity and better all-round performance. This emphasis upon performance and hierarchisation of everyday practices has consequences for those practices. In their study of classification systems, Geoffrey Bowker and Susan Leigh Star argue that classification ties 'the person into an infrastructure—into a set of work practices, beliefs, narratives, and organisational routines' (1999, p. 319). We can thus ask, what effect does counting steps have upon an everyday practice such as walking? What relationship of the self to the body (to indulge in such a dichotomy for the moment) does this produce? The disciplinary aim to generate longitudinal data, which is produced by activity across time, means that users compete with themselves and others in social networks to increase or maintain their daily step count. Walking becomes a goal and an object of accumulation. This objectification of walking as an activity works in stark contrast to Frederic Gross' claim in *Philosophy of Walking* that you are 'doing walking when you walk, nothing but walking. But having nothing to do but walk makes it possible to recover the pure sensation of being, to rediscover the simple joy of existing, the joy that permeates the whole of childhood' (2014, p. 83). Gross' account may be romantic, but it demonstrates that the process of objectification reduces the body to a kind of machine. Sociotechnical measurement machines submit the flesh of the body—and its corporeality—to an instrumental order. Walking is no longer simply for walking, it is reconfigured as an activity for something else, for enhanced performance.

A similar case about changing perceptions can be made for eating. Health motivation apps encourage healthy diets by providing information on the nutritional aspects of food. Users can take control of their diet by tracking calorie intake as well as considering the nutritional contents of specific foods. Food is effectively reduced to its nutritional elements, and eating becomes a calculable input within bioinformationalised everyday life.[1] Rather than the pleasure and culture of food, eating and the body are reduced to communication and data processes. With this process, food simply becomes a means for responding to the body as a dynamic system rather than a pleasure in itself. Subjects simply eat to offset the risk of disease and to optimise performance levels. The nutritional recommendations of health motivation apps perpetuate what Gyorgy Scrinis calls an ideology of 'nutritionism' (2008). Nutritionism is the food science version of dataism. Scrinis explains,

> The nutritionism paradigm ... is defined by an overly reductive focus on this nutri-biochemical level. Particular nutrients, food components, or biomarkers—such as saturated fats, kilojoules, the glycemic index (gi), and the body mass index (bmi)—are abstracted out of the context of foods, diets, and bodily processes. Removed from their broader cultural and ecological ambits, they come to represent the definitive truth about the relationship between food and bodily health. Within the nutritionism paradigm, this nutri-biochemical level of knowledge is not used merely to inform and complement but instead tends to displace and undermine food-level knowledge, as well as other ways of understanding the relationship between food and the body. (2008, p. 40)

The health motivation app makes food recommendations in terms of the nutritionism paradigm. The following typical message from the UP system app situates the body in relation to food elements in precisely these terms:

> Get your kick. Only 56% of Americans reach the recommended daily potassium goal of 4,700mg. The mineral lowers blood pressure and reduces risk of osteoporosis. Add 1 cup of spinach to dinner tonight for a 740mg kick. (UP system app)

The message works within a four-part arrangement that requires the modulation of practices. Firstly, the user is located within a broad context: potassium is an essential mineral and many Americans do not take in enough. Users outside the United States might understand this as "potassium is important and it is common for levels to be low". Secondly, users learn about the health benefits of potassium, articulated in the language of risk: eating potassium-laden foods reduces the risk of osteoporosis. Thirdly, users are encouraged to reconfigure their evening meal in terms of information about this trace element. What counts is the scientific precision signified by the numerical value '740mg' rather than how spinach tastes. This exact measurement, compared to a less exact statement such as 'high in potassium', signifies scientific precision and authority. Interestingly, the message does not offer alternatives to spinach either: its availability is assumed. Finally, the message is framed via a colloquial term: 'kick'. Users cannot directly experience the potassium in spinach, but the term implies that the consumption of spinach leads to a natural high and increased levels of potency. 'Kick' is derived from the language of marketing: spinach is rendered as a commodity that benefits consumers in specific ways. Health app users become, as Richard Sennett contends in his analysis of consumer capitalism, 'consumers of potency'. Consuming commodities in contemporary societies is reinforced by the 'promise of expanding our capability' (2006, p. 154). Spinach offers the promise of potency. As can be seen, the message encourages and defines a practice, informing users why they should eat spinach and what happens when they do.

Another typical example of the modulation of practices comes in the form of lifestyle tips. Many health motivation apps include tips drawn from lifestyle websites. One such message titled 'Get Smooched' informs users, 'Kissing is good for headaches, high-blood pressure and even cavities. Your mouth secretes saliva when you smooch, washing away nasty dental plaque. Pucker up!' (UP system app). This example may be trivial, but what it reveals is the medicalisation of social practices. The pleasure of kissing is reconfigured here as a healthy practice.

From the perspective of health motivation app users, the power and persuasiveness of this medical information is immediately apparent. The capacity to generate, record, and communicate fundamental personal data allows subjects to use numbers and graphs as a unique marker of identity. The data allows subjects to discover information about themselves. The health motivation app effectively takes an impersonal form of knowledge construction—data—and personalises it. Subjects produce data through their daily practices, and then this data becomes an intimate object. This is me! The personalised data signifies the presence of subjectivity. This presence is then validated by generated messages such as 'Well done. 3 Day streak averaging 12,154 steps per day. Since you're cruising, try to reach your goal 2 more days for a 5 day streak' (UP system app).

Alongside the graphical information, the app also typically employs a decision support system in the form of individually tailored recommendations. Such recommendations include an optimum time for going to sleep at night, to increase the daily step goal, to cut down on salt intake, and so on. For example:

> This past week, your longest idle time was 1hr 22m on Tuesday. If you want to move, take some extra loops around the block, or stand up now and do a little dance. (UP system app)

This recommendation system is produced through statistical analysis. As I stated earlier, the app makes correlations between its key indicators—steps, sleep, food, mood—and then produces the tailored recommendations and encouragements. Crucially, the recommendations that are offered to the user are an unknown form of knowledge. In other words, the user finds hidden knowledge about themselves through the use of the app. It is in this sense that statistical reasoning prevails. The algorithm that processes the data is not transparent. The app is part of the larger process of the bioinformationalisation of life. That is, users are required to submit to a non-transparent statistical machine that modulates everyday life in particular and ever-modifiable ways.

Conclusion: Dataism to Control

The exponential uptake of wearable health technologies suggests that the process of bioinformationalisation is already well-entrenched in everyday life. We can thus return to consider how health motivation apps connect everyday life to the processes of governance and control. At the most rudimentary level, we can deduce that health motivation technologies are entangled with the prevailing view that in techno-capitalist societies it is difficult to maintain a healthy lifestyle. Health app users must, therefore, be vigilant; their everyday lives should be constructed around disciplined routines and practices. There are, no doubt, many reasons why such vigilance is a requirement for health. The prevalence of cheap fast food and the sedentary life that marks techno-capitalist societies are two reasons (Virilio 2000, pp. 71–87). The associated 'obesity epidemic' is now taken as a given by governments, media, and health care systems across the world (Seidell and Halberstadt 2015). As Lauren Berlant puts it, this 'epidemic' has been met by the state with 'slow-death crisis-scandal management' strategies (2011, p. 762). Health motivation apps, it would seem, are at the forefront of combating this epidemic from the perspective of governance. The government-mandated potassium consumption levels, discussed above, bear this out.

Alongside governance, corporations also champion wearable technology as a means for combating the problem of health and increasing worker productivity. As Aviva Rutkin notes,

> [m]any companies—including BP, eBay and Buffer—already encourage employees to wear activity trackers like the Fitbit, often in exchange for discounts on health insurance … In cases like these, wearables are designed to boost the health and general productivity of the employees, sometimes encouraging them to compete against one another online. That makes sense: a healthier workforce saves a company money in the long run. (2014, p. 22)

Rutkin's account of the value of wearable technology for corporations is cause for concern. Here we find a strange situation in which workers compete against each other to see who can be the healthiest, all for the benefit of the corporation. There is no doubt that wearable technology development is being shaped by corporate interests. As William Davies points out, 'employers, health insurers and wellness service providers are amongst the main enthusiasts for the [i]phone's constant measurement of bodily behavior' (2015, p. 135).

This enthusiasm for wearable technology signals new possibilities for reconfiguring the relationship between technology, workers, and capital. Karl Marx signalled in a much-cited passage on machines in the *Grundrisse* (1993) that technologies alter the relationship between workers and capital. With large scale industry, Marx predicted, 'the creation of wealth comes to depend less on labour time and on the amount of labour employed than on the power of the agencies set in motion during labour time' (1993, p. 704). The result of this scenario is the increasing subsumption of the subject by the capitalist machinery, with the rhythm of everyday life being altered to the time of the machines. '*The most developed machinery*', Marx writes '*forces the worker to work longer ... than he did with the simplest, crudest tools*' (pp. 708–709; original emphasis). The whole of social life thus becomes increasingly ordered by capital. This machinic modernity, of course, alters once again with the invention of new technology for the production of wealth. If Marx was concerned with the application of science to production, we should today consider the application of 'communication science', broadly understood, to the rhythm of everyday life. We can point, as Antonio Negri argues, to a dispersed capitalism that works though networks and real-time communication systems as evidence for this claim. Wearable technology users are a recent example of what Negri calls the 'socialised worker'. He writes, 'communication is to the socialised worker what the wage relationship was to the mass worker' (1989, p. 118).[2]

Drawing on Deleuze's account of control societies, Negri makes the case that capitalism now depends upon communicative systems to develop and maintain growth. This opens up a new terrain of struggle. Communicative processes and generated information and statistical models form, for Negri, an antagonistic relation between workers and capital. Capitalism works, as Nick Dyer-Witheford explains, by 'furnishing and familiarizing labor with a "wired" habitat through which instructions can be streamed and feedback channelled' (1999, p. 85). Channelled feedback is captured and packaged in order to maximise efficiencies and generate profit. Herein lies the crux of Negri's argument: the creative, communicated activity of workers and users generates value. This is precisely how FitBit, Jawbone, and others can become profitable. Generated user data has commercial value in the context of marketing, health commodities, and governance. However, data also poses a problem. It is difficult to make data meaningful in useful and profitable ways. Moreover, users modulate behaviour, but such modulations can open up unforeseen consequences, and it is possible for users to network and produce data for purposes outside or in opposition to corporate interests. This is precisely what Negri means when he argues that there is a tension between communication and information. With respect to health motivation apps, however, it remains to be seen how the resistive uses of communicative systems will manifest.

What we see currently, as a consequence of this tension, is an increase in data input points. Data is crucial in the control society context. Control societies consist of communication systems generating data via input from sensors. Accurate data enables effective decision-making, efficient production, marketing, and profit. The predictive capacity of data is directly proportional to the number of sensors in the field, and wearable technologies multiply the number of these sensors. This is how sociotechnical measurement machines such as health motivation apps work in the context of societies of control. Crucially, everyday life involves an interlinked relationship between user and technology. Users live with and through the technology (thus the term "user" is problematic). The apps connect bodies to a network of sensors that feed data into an algorithm. The algorithm produces information about user activity which is then returned back to the user in the form of recommendations. Thus a newly formed somatechnics of the sensored body is beginning to emerge, in which everyday activities modulate in accordance with the modulations of the control society, and in which the body is produced within a security-based governmentality epitomised in the ideology of dataism. The contemporary, somatechnically instrumentalised body is increasingly indissociable from the sensors that measure it, the data that describes it, and the control modulations that govern it.

However, we must not lose sight of the term 'control' as Deleuze uses it in its strict sense: to control is to keep the wayward on track. This means that sensored bodies are unruly from the perspective of power. As the tension between capital and communication identified by Negri suggests, the sensored body is awkwardly situated within the contemporary capitalist formation. The health motivation app thus marks several tensions within capitalist societies. Firstly, the aforementioned task of implementing app technology—software updates, recharging armbands, reconfiguring the gait, etc—involves levels of technological and physical competency that are not even across populations. All bodies are not the same. This unevenness limits the scope of the apps' integration with everyday life. Secondly, the apps locate the tension within capitalism between the consuming and the productive body. To stretch Sigmund Freud (1922), health motivation apps in effect confront the pleasure principle with the reality principle. The sacrificial, even puritan, "pain makes gain" mantra that characterises health and fitness is directly at odds with the comforts and indulgences that capitalism promises as a reward for labour. Finally, capitalism contradictorily produces the unhealthy and individualistic consuming subject that it then seeks to control for the purposes of production. Modulating subjects are effectively caught between two competing trajectories: one moving toward consumption and sedentariness, and the other toward healthy mobility, productivity, and accumulation. Users can follow the health regime closely and maintain high levels of security, but within capitalist culture the pleasure principle can never be completely overcome. As a consequence, pleasure is instrumentalised as a somatechnic of aspiration in relation to the vigilance and discipline required to be healthy. At the same time, refusing to submit to this regimentation can be a valued and desirable expression of individuality and agency, signalling that the somatechnic reconfiguration of everyday life is fraught and unpredictable.

Health apps and wearable technologies are at the centre of these three tensions: the confidence (and anxiety) placed in data versus the effort and competence required to make it accurate, the displeasure of fitness and healthy eating versus consumerist indulgence, and the production of individualistic and aspirational subjects versus manipulable drones. Health motivation technologies thus occupy contradictory social and cultural spaces within contemporary capitalism, and their modulation from machines of individual freedom to machines of biopolitical regimentation is reflected in their dual functions of personalising and individuating data, while simultaneously implementing it in broader processes of security and population.

Notes

1. See Joseph Pugliese's contribution in this volume for an account of bioinformationalisation.
2. See also Sy Taffel's contribution in this volume on Deleuzian control over labour in Amazon warehouses.

References

Amazon. (2013). Amazon.com: TheChamp's review of UP24 by Jawbone Activity Tracker. http://www.amazon.com/review/R3G3VGVEJZ730Z. Accessed 1 Oct 2014.

Andrejevic, M. (2009). Privacy, exploitation, and the digital enclosure. *Amsterdam Law Forum, 1*(4), 47–62.

Ayo, N. (2012). Understanding health promotion in a neoliberal climate and the making of health conscious citizens. *Critical Public Health, 22*(1), 99–105.

Azumio. (2015a). ARGUS by Azumio. http://www.azumio.com/s/argus/index.html. Accessed 7 Sept 2015.

Azumio. (2015b). Contact Azumio. http://www.azumio.com/s/contact/index.html. Accessed 7 Sept 2015.

Berlant, L. (2011). Slow death (obesity, sovereignty, lateral agency). *Critical Inquiry, 33*(4), 754–780.

Bowker, G., & Star, S. L. (1999). *Sorting things out: Classification and its consequences.* Cambridge: MIT Press.

Davies, W. (2015). *The happiness industry: How the government and big business sold us well-being.* London: Verso.

Deleuze, G. (1992). Postscript on societies of control. *October, 59,* 3–7.

Desrosières, A. (1998). *The politics of large numbers: A history of statistical reasoning.* Cambridge: Harvard University Press.

Dyer-Witheford, N. (1999). *Cyber-Marx: Cycles and circuits of struggle in high technology capitalism.* Urbana: University of Illinois Press.

Fitbit. (2015). About Fitbit. http://www.fitbit.com/nz/about. Accessed 7 Sept 2015.

Foucault, M. (1977). *Discipline and punish: The birth of the prison.* New York: Pantheon Books.

Foucault, M. (1990). *The history of sexuality: An introduction, volume 1*. New York: Vintage Books.

Foucault, M. (2009). *Security, territory, population: Lectures at the Collège de France, 1977–78*. Basingstoke: Palgrave Macmillan.

Freud, S. (1922). *Beyond the pleasure principle*. London: International Psycho-Analytical Press.

Gordon, C. (1991). Governmental rationality: An introduction. In G. Burchell & C. Gordon (Eds.), *The Foucault effect: Studies in governmentality*. London: Harvester Wheatsheaf.

Gray, D. (2014). DSB review: Jawbone UP24. 25 February. http://dorishinyblog.com/2014/02/dsb-review-jawbone-up24/. Accessed 1 Oct 2014.

Gross, F. (2014). *A philosophy of walking*. London: Verso.

Hall, C., Figuero, A., Fernhal, B., & Kanaley, J. A. (2004). Energy expenditure of walking and running: Comparison with prediction equations. *Medicine & Science in Sports & Exercise, 36*(12), 2128–2134.

Jawbone. (2015a). UP by Jawbone | A smarter activity tracker for a fitter you. https://jawbone.com/up. Accessed 7 Sept 2015.

Jawbone. (2015b). Jawbone | About the company. https://jawbone.com/about. Accessed 7 Sept 2015.

Lupton, D. (2014). Critical perspectives on digital health technologies. *Sociology Compass, 8*(12), 1344–1359.

Mandl, K., Mandel, J., & Kohane, I. (2015). Driving innovation in health systems through an apps-based information economy. *Cell Systems, 1*, 9–13.

Marx, K. (1993). *The grundrisse*. London: Penguin.

Negri, A. (1989). *The politics of subversion*. Cambridge: Polity Press.

Pugliese, J., & Stryker, S. (2009). The somatechnics of race and whiteness. *Social Semiotics, 19*(1), 1–8.

Pugliese, J. (2010). *Biometrics: Bodies, technologies, biopolitics*. London: Routledge.

Quantified Self. (2015). About the quantified self. http://quantifiedself.com/about/. Accessed 1 Sept 2015.

Research2guidance. (2014). mHealth app developer economics. http://mhealtheconomics.com/mhealth-developer-economics-report/. Accessed 9 Sept 2015.

Rutkin, A. (2014). Off the clock, on the record. *New Scientist, 224*(2991), 22–23.

Scrinis, G. (2008). On the ideology of nutritionism. *Gastronomica: The Journal of Food and Culture, 8*(1), 39–48.

Seidell, J., & Halberstadt, J. (2015). The global burden of obesity and the challenges of prevention. *Annals of Nutrition & Metabolism, 66*(2), 7–12.

Sennett, R. (2006). *The culture of the new capitalism.* New Haven: Yale University Press.

Swan, M. (2013). The quantified self: Fundamental disruption in big data science and biological discovery. *Big Data, 1*(2), 85–99.

van Dijck, J. (2014). Datafication, dataism and dataveillance: Big data between scientific paradigm and ideology. *Surveillance & Society, 12*(2), 197–208.

Virilio, P. (2000). *Polar inertia.* London: Sage.

7

Invisible Bodies and Forgotten Spaces: Materiality, Toxicity, and Labour in Digital Ecologies

Sy Taffel

From the radioactive waste associated with mining rare earth elements to the workers who earn US$1 a day treating highly toxic e-waste, behind the image of digital culture as "green" and "weightless", we find numerous bodily harms enacted upon humans and nonhumans overwhelmingly located in spaces far removed from the primary sites of technological consumption. This chapter explores a range of social and ecological costs attributable to the life-cycle of digital technologies, particularly focusing upon four specific places and moments within the life-cycle of microelectronics: the extraction of rare earth elements from Baotou, China; the manufacture of computational devices within Chinese factories; Amazon's European distribution warehouses; and artisanal e-waste recycling in Guiyu, China. These sites have been chosen as they collectively foreground a range of deleterious social and environmental impacts associated with digital technologies. Situating microelectronics within life-cycles and other ecological processes in this manner is designed to remove the focus

S. Taffel (✉)
Media studies, Massey University, Aotearoa, New Zealand

© The Author(s) 2016   **121**
H. Randell-Moon, R. Tippet (eds.), *Security, Race, Biopower*,
DOI 10.1057/978-1-137-55408-6_7

upon digital technologies as short-lived objects of desire—instead asking us to re-think the material politics of digital culture.

The material architecture which underpins the digital revolution is commonly referred to in terms which postulate only the most tenuous of connections to the hardware layer of contemporary digital ecosystems. Obscured behind the language of "virtual" reality, "immaterial" labour, "cognitive" capitalism, and "cloud" computing which positions digital technoculture as "smart", "green", and "weightless", the material impacts associated with various stages of the life-cycle associated with digital technologies often becomes invisible. Indeed, whilst the intersection between bodies, spaces, and technologies has long been considered pivotal to comprehending the social changes wrought by digital culture, the focus of such analyses has tended to focus overwhelmingly upon sites of technological consumption and the effects of digital technologies upon end users.

Dating back to the mid-1980s and Donna Haraway's (1991) *Manifesto for Cyborgs*, scholarly approaches have emerged, which investigate the ways contemporary technoscientific endeavours have created a proliferation of hybrid entities—which Haraway (2003) terms 'naturecultures'—denoting that it has become impossible to remove the technocultural elements from any notional human "nature". However, whereas Haraway's manifesto was an ironic myth aiming to decentre the anthropocentric focus of cultural studies and reject eco-feminist discourse which identified women with nature, the dominant technoscientific framework of understanding humans, animals, machines, and other entities through systems theory (Kelly 1994) reduces them to information processing nodes within networks governed by an informatics of domination; a cybernetic system of power intent on reducing the diversity of existence to a single, allegedly post-political language of information. Such claims for a universal, value-free language of objective rationality are understandably critiqued by feminists such as Haraway, N. Katherine Hayles (1999), and Rosi Braidotti (2013) for whom such rhetoric has always been the ideological veil of phallocentric bourgeois discourse, whose hegemonic social status becomes projected as the universal project of the "(hu)man" from which the project of critical posthumanism seeks to disentangle itself.

Whereas cyborgian discourses focused upon hybridised biotic and artificial systems, recent work around neuroplasticity and technology (Damasio 2008; Malabou 2008; Terranova 2012) has used developments in cognitive science, especially those pertaining to the extent of the plasticity of the human brain (Pascual-Leone and Torres 1993; Rose and Abi-Rached 2013; Gindrat et al. 2014), to contend that humans have always been cyborgs. As the development of neural pathways depends upon the interaction of the somasensory systems with an exterior technological and cultural environment, we can follow Bernard Stiegler's (1998) claim that being human necessarily involves a technocultural dimension. This ontological dimension moves away from a static sense of being and human nature towards comprehending technocultural development and evolution as dynamic drivers of biological changes within the human elements of an entangled sociotechnical milieu, whilst human intentionality simultaneously impacts upon those technocultural forms creating the complex feedback-driven circuitry of epigenetic technocultural evolution.

Whilst posthuman, cyborgian, and neurological approaches to media, space, and bodies ask pertinent questions surrounding the current computational technocultural milieu, their focus is predominantly upon the end users of technologies, either as consumers or as immaterial workers (Lazzarato 1996) for whom information and communication technologies (ICTs) are essential tools for contemporary biopolitical production (Hardt and Negri 2004, pp. 93–95). Spatially, this entails concentrating upon the urban and domestic sites of technological consumption, closely corresponding to what Manuel Castells (1996, pp. 407–453) has termed the 'space of flows' within a globalised network society. The space of flows is largely dominated by the preferentially attached nodes formed by urban and financial centres, within which space is increasingly collaboratively constituted and transduced by computational assemblages, forming a dyadic relationship between digital technologies and space which geographers Rob Kitchin and Martin Dodge (2011) have termed 'Code/Space'. Code/Spaces allow for speeds and volumes of human and nonhuman traffic which exceed the capacities of previous technocultural assemblages, whilst affording further concentrations of wealth and power within these spaces.

However, the construction of contemporary Code/Spaces is predicated upon a meshwork of global flows of materials, labour, and capital which have deleterious impacts upon the bodies of human and nonhuman actors whose spatial distributions depart from the polycentric model of the space of flows. Echoing Mark Weiser's (1991) often repeated mantra that the most profound technologies are those which effectively disappear from public consciousness, simply becoming part of the fabric of everyday life,[1] this chapter seeks to draw attention towards the spaces and bodies affected by the construction of the material infrastructure of digital culture, but which are often invisible to the publics who consume digital technologies. Foregrounding the deleterious impacts associated with the forgotten spaces required for the construction of digital technoculture thus forms a line of technological critique which contextualises the allegedly post-political and non-ideological positioning of ICTs within the material practices of a globalised capitalism implicated in increasing levels of social inequality and the commodification of an expanding range of communicative practices (Dean 2009; Fuchs 2010). Importantly, as Joseph Pugliese (2007) and Sean Cubitt (2014) have shown in very different contexts, this inequitable technoculture has a thoroughly postcolonial and racial dynamic to it, which frequently becomes hidden due to the supposedly neutral and objective functioning of mathematical technology that masks the cultural and corporeal biases programmed into computational hardware.

The focus upon flows of material and energy through multi-scalar, spatio-temporal systems which range from the microscopic transistors used in CPUs to global networks of fibre-optics, from the deep time (Zielinski 2006) of geological mineral formation to the speed of light transmission of electronic data, places this approach within the methodological nexus of media ecology (Fuller 2005; Parikka 2011; Taffel 2013). Homologous to the way that ecologists trace flows of energy through ecosystems containing humans, other biotic systems, and non-living elements, media ecologists approach flows of materials and energy within media systems comprising humans and nonhumans. This approach departs from methods predicated upon reducing media to representations, resonating with the affective approach of non-representational theory (Thrift 2008) and the broader theoretical currents of new materialism (Coole and Frost 2010; Parikka 2012) which emphasise media as material processes.

Extracting Rare Earths

Thinking in ecological terms requires proceeding in terms of cycles, loops, and flows rather than positing origins or beginnings. Indeed, examining the materials required for microelectronics in this manner necessitates attention to the formation of the various depositions of minerals, metals, fossil fuels, and other substances required for digital technologies (Parikka 2015). However, in considering the "life-cycle" of microelectronics, we can adjust our temporal focus from geological formation to the process of material extraction, the anthropogenic removal of the metals, minerals, and ores required for microelectronics manufacture. This does not posit extraction as the point of origin, but a particular moment within the deep time of media processes, which is useful for considering the embodied politics that exist within digital devices, bodies, and places. Whilst a proportion of materials are obtained from recycled microelectronics, most devices are not currently recycled (Electronics Take-Back Coalition 2011). Those which are recycled only re-use particular valuable substances rather than the entire assemblage of materials. Producing digital infrastructure, then, requires the removal of vast amounts of material from the crust of the Earth, connecting supposedly smart and green ICTs to the literally and metaphorically dirty networks of the global extraction industries.

An example which foregrounds the embodied and geopolitical impacts of the extraction and refinement processes associated with ICTs arises from the production of Rare Earth Elements (REE), a set comprised of seventeen elements containing the fifteen lanthanides, yttrium, and scandium. Whilst their collective title suggests scarcity, this is somewhat of a misnomer, with cerium being more abundant than copper and several other REE such as yttrium, neodymium, and lanthanum being more common than cobalt, lead, and lithium. However, REE frequently occur in extremely low concentrations and are combined together within ores, so extracting individual REE is complex due to their similar chemical composition, and consequently deposits are often not commercially exploitable.

Despite the complexities involved in procurement, REE are integral to numerous ICTs such as hard disk drives, microphones, headphones, display technologies, and fibre-optics. In each case, the properties of the specific REE are essential to the function of the digital devices. Neodymium magnets are the strongest type of permanent magnet (Fraden 2010, p. 73), affording the construction of powerful, permanent magnets which are relatively small and lightweight, which comprise a key innovation surrounding the miniaturisation of microelectronics and audio-visual equipment (Haxel et al. 2005). Europium is primarily used in phosphors, materials which emit wavelengths of visible light when exposed to an electron or light source, and historically was used to create the reds in colour cathode ray tube television sets. Contemporary display-related technologies which continue to utilise europium as a phosphor include liquid crystal displays (LCDs), light emitting diodes (LEDs), and plasma display panels (Schulze 2014). Despite costing up to US$1700/kg (Haxel et al. 2005), europium's unique properties as a phosphor continue to make it an attractive material within ICT production. Erbium is the most commonly employed optical fibre amplifier within opto-electronics because 'its wavelength amplification window coincides with the third transmission window of silica-based optical fiber' (Sharma et al. 2013). Erbium's specific affordances are ideally suited for use in the fibre optic cables whose bandwidth has been pivotal to the formation of media-rich online environments. Considering these usages of REE together reveals their importance to the production of audio, visual, storage, and transmission technologies associated with digital technoculture. Without the availability of REE, the contemporary micro and opto-electronics ecology inhabited by YouTube, Facebook, and other social media would be significantly transformed.

From 1965 until the mid-1980s, the USA was the major global source of REE, with a single mine located at Mountain Pass in California responsible for the majority of global production. From the mid-1980s onwards, China became the major player in the industry, producing REE at far cheaper prices than the Mountain Pass operation, which consequently closed in 2002. By 2012, China was responsible for 95 percent of global REE supplies (Ives 2013). Chinese control over crucial components of digital infrastructure has caused concern amongst other nation states (particularly the USA and Japan), especially following a decision taken by the Chinese government in 2009 to produce export taxes and quotas for REE. In 2012, President Obama announced that the USA, along with Japan and the European Union (EU), had filed a case with the World Trade Organisation (WTO) over these policies, which he argues were designed to provide a competitive advantage for the domestic Chinese market (Humphries 2013, p. 18). Partially as a consequence of these geopolitical issues associated with concern over hegemonic Chinese REE production and the resultant uncertainty introduced into global mineral markets, numerous alternative production sites have come online since 2012, including the reopened Mountain Pass mine. Consequently, the Chinese percentage of the REE market has fallen from 95 percent to 85 percent (Kaiman 2014).

Interestingly, the reason the Chinese government provided the WTO for introducing export taxes and quotas was that money was needed to fund a clean up operation due to the environmental degradation caused by REE mining:

> According to the Chinese Society of Rare Earths, 9,600 to 12,000 cubic meters (340,000 to 420,000 cubic feet) of waste gas—containing dust concentrate, hydrofluoric acid, sulfur dioxide, and sulfuric acid—are released with every ton of rare metals that are mined. Approximately 75 cubic meters (2,600 cubic feet) of acidic wastewater, plus about a ton of radioactive waste residue are also produced. (NASA 2012)

The radioactive water is produced because REE deposits tend to occur alongside radioactive elements such as thorium and radium, which are consequently present in the wastewater tailings associated with REE mining. The acids and sulphates are required for the process of separating out the valuable REE from one another and other substances.

China's largest REE facility, located at Bayan Obo, has a tailings pond 11 square kilometres in size, which holds an estimated 180 million tonnes of waste from the mine. Since 2009, Baotou Steel, who own the mine, have been relocating farmers away from the pond following spikes in cancers amongst the human population and growth defects amongst livestock (Kaiman 2014). Such effects on humans and nonhumans proximate to extraction industries are a stark reminder that beneath the marketing of ICTs as smart and green, lie commodities whose production is entangled with ecologically devastating industry.

It is worth noting that the detrimental ecological impact arises from the immense volume of hazardous waste generated in extracting and separating the materials. Conventionally we associate waste with disposing of the final commodity, or we reassure ourselves that by recycling materials we are avoiding waste entirely. In fact, REE reveals that the production of waste is present at multiple stages within the life-cycle of ICTs, and that in many cases, the volume of waste generated in production is far greater than that of the final product, and can have devastating impacts upon living systems located close by. Indeed, to manufacture a 129 gram smartphone such as Apple's iPhone 6 requires around 75 kg of raw materials (IEEE 2013), entailing that 99.9 percent of the volume of materials used are not seen in the final device.

Manufacturing Misery

The most recent, widely publicised case of labour rights abuses in microelectronics manufacturing came from a British Broadcasting Corporation (BBC) Panorama television programme aired in December 2014, which focused upon conditions at a Pegatron factory in China that produces iPhones and iPads for Apple (Bilton 2014). Employees regularly worked in excess of both the 49 hours stipulated as the maximum by Chinese law and the 60 hours permitted by Apple's guidelines—and during busy periods, workers laboured for 18 consecutive days with requests for a day off being routinely denied. This occurred due to factory operators signing contracts which set quotas of devices required for specific dates (usually relating to the global launch of a new model), and which mandate severe financial penalties for the subcontractor for any shortfall in production. In the recording of the Panorama documentary *Apple's Broken Promises*, workers were filmed falling asleep during their shifts, their exhausted bodies unable to maintain the level of productivity demanded by the global microelectronics industry (Agence France-Presse 2014). The images of drained and weary bodies struggling to stay awake during the gruelling shifts assembling cutting edge consumer microelectonics further undermines the pretence that digital technologies are predicated upon "immaterial" labour. The Panorama documentary was not, however, the first revelation that labour undertaken by subcontracted workers in Chinese microelectronics factories frequently breaches local labour laws and the codes of conduct which corporations such as Apple hold up as evidence of their corporate social responsibility.

The first major microelectronics scandal arose following a 2006 story published in the United Kingdom-based *Mail on Sunday*, uncovering the labour and living conditions of the Chinese workers who assembled Apple's iPods at Foxconn's plant in Longnua, Shenzhen. Employees were found to work for up to 15 hours a day, with large volumes of compulsory overtime being effected by management at peak times. Workers at the 200,000-person complex were housed in dormitories with up to a hundred occupants each. Some plants were found to charge the workers almost half their meagre earnings for rent and food in the on-site dormitories and cafeterias (Joseph 2006). The *Mail on Sunday* article contrasted the cost of an iPod in the UK, between £109 and £179, with those of the parts—around £41—and the £4.20 in labour costs as a way of highlighting the deeply inequitable situation surrounding the globalisation of the microelectronics industry: corporations such as Apple make vast profits off the backs of impoverished Chinese factory workers.

Perhaps the most notorious case surrounding labour and microelectronics manufacturing again pertains to Foxconn's Longnua facility, which made global headlines when fourteen young migrant workers committed suicide within a two-month period during 2010. For these young workers, the practice of creating the architecture for a supposedly democratising and empowering participatory digital culture was the source of such immense immiseration that suicide became a means to escape the long hours, forced overtime, and fierce disciplinary regime in place. Indeed, during July 2009, Sun Danyong, another Foxconn employee, committed suicide after reporting that a prototype iPhone 4 in his care had been lost. Danyong reported that consequent to filing his report, his residence was searched and he was interrogated and beaten by Foxconn security (Barboza 2009).

In addition to labour rights abuses, there have been instances where toxic materials involved in microelectronics manufacturing have caused serious injury and illness to factory workers. In 2009, Wintek—another subcontractor working for Apple in China, who manufactures LCD and touchscreen display panels for consumer devices such as the iPhone—opted to use n-hexane rather than alcohol to clean panels, as it dried more quickly and was less prone to leaving streaks (Chen 2010). N-hexane, however, is a potent neurotoxin, which causes damage to both the central and peripheral nervous systems in humans (Liu et al. 2012). The situation caused particularly acute harm because the dangers of adopting the substance had not been adequately assessed, and consequently, n-hexane was used without sufficient ventilation to remove the vapours. Apple's 2011 annual review of labour conditions at its subcontractor's facilities revealed that 137 workers at the Wintek plant had been seriously injured by n-hexane poisoning (Barboza 2011), with many individuals being hospitalised for up to several months with nerve damage that manifested as symptoms such as headaches, dizziness, loss of the sense of touch, weakness, numbness, limb pain, and extreme fatigue. In interviews, workers complained that they were so enfeebled by n-hexane poisoning that they would fall over if someone touched them (Branigan 2010), or were so hypersensitive to cold temperatures that they had to wear insulated clothing at all times (Barboza 2011).

As with the case of material extraction, we find that microelectronics manufacturing contains instances whereby serious harms are enacted to the bodies of those involved in parts of the product life-cycle which are spatially far removed from the predominant sites of consumption. Globalisation here again involves a process of externalising costs, whereby brand name corporations use cheap subcontracted labour with inadequate safety programmes and draconian disciplinary measures to lower operating costs and improve profit margins at the expense of the labourers.

In each case involving Apple, the company issued press releases affirming their dedication to improving labour conditions, positing Apple as an ethically motivated corporate actor. These public relations exercises and the programmes they vociferously point toward are severely undermined by repeated exposure of ongoing health and labour violations with subcontractors. If Apple took the measures they outlined in 2005, such as ensuring that workers did not work significantly over the legally permitted number of weekly hours, the same issues would not remain widespread in 2014, when the BBC's Panorama exposé aired. Corporations take low-cost contracts which can deliver cut-price devices because of these labour rights and health and safety abuses. Contrary to the claims of corporate social responsibility and public relations departments, as Milton Friedman (1970) famously stated, 'There is one and only one social responsibility of business—to use its resources and engage in activities designed to increase its profits.' If Apple genuinely wished to transform labour relations, they could purchase manufacturing facilities outright and implement higher wages, reduce working hours, and improve health and safety practices. Doing so, however, would impact upon the profit margins of one of the planet's most valuable corporations.

Cybernetic Distribution

A huge amount of ICT equipment goes towards the server farms, cell phone tower networks, underwater fibre-optic cable relays and other back-end infrastructure enablers of the thin-client access systems, which allow end consumers to use mobile computational devices such as smartphones and tablets to access and process a variety of data forms in wireless Code/Spaces. However, these access devices, along with a multitude of accessories and components, go into the international consumer sales market. When it comes to online retail, the dominant player in much of the world is Amazon.com, which launched in 1994 as an online bookstore, but rapidly diversified towards selling a vast array of goods, and currently positions itself as a dominant player not only in online retail, but in cloud storage and hosting services, where it sees its competitors as Google and Facebook rather than other retailers.

Amazon's gargantuan warehouses are often over 250,000 square meters in size, and utilise a cataloguing system known as chaotic storage (Sterling 2013). Unlike traditional cataloguing and storage systems, which sort items into homogenous sections, chaotic storage sees delivery workers simply place items onto any unoccupied shelving. The items and shelves both contain barcodes which are scanned, so the warehouse database knows the location of the goods. Once an order is placed, pickers are dispatched to the items along the shortest possible route through the warehouse, saving a significant amount of time in comparison to traditional storage systems, as chaotic distribution prevents pickers having to traverse the entire warehouse. Such a cybernetic system with constantly modulating picker routes predicated upon varying stock levels and placement exemplifies the Deleuzian logic of control (Deleuze 1992), which emphasises fluidity and dynamism enabled by assemblages of computational and human labour. This database-driven system also demarcates the Amazon warehouse as a Code/Space, a geography which would effectively fail to function were it not for the database and algorithms which underpin chaotic storage.

This drive towards cybernetic efficiency and the consequent reduction in labour costs is equally present within Amazon's ethically dubious warehouse labour practices. Amazon employs an extreme version of Taylorism, whereby staff who are frequently employed on temporary, "zero-hour" contracts are given tasks which are measured in seconds by Global Positioning System (GPS)-based scanning devices. These devices dictate the precise route workers must follow through the warehouse, and if workers fail to meet their allotted quota of tasks, they receive warnings either by text message or in person, with a rigid three-strikes-and-you're-out policy in place. Indeed, workers have been fired by Amazon for being found guilty of spending several unproductive minutes during a 10-hour shift (Head 2014, p. 40). Often, pickers must walk 20–25 kilometres per shift, entailing that elderly, unfit, or disabled workers are soon unable to keep up with individualised targets, which regularly increase over time in accordance with past performance based upon GPS-derived data. In the UK, where labour laws entail that after twelve weeks employees receive additional rights, Amazon regularly fires staff after eleven weeks and subsequently re-hires them, thereby avoiding having to offer statutory rights such as holiday pay (Bilton 2013). These workers receive the minimum wage allowed by each state, i.e. less than a living wage, for highly precarious labour, delineating that even within Organisation for Economic Co-operation and Development (OECD) nations such as the USA, UK, and Germany, there are serious issues surrounding labour practices within the microelectronics life-cycle.

Additionally, Amazon demonstrates the way that the precise quantification of movement afforded by pervasive media technologies such as GPS can be utilised to enact forms of disciplinarity which were inconceivable under previous technological regimes. Whereas the Foucauldian notion of discipline was predicated upon subjects internalising the possibility of surveillance at any given moment and thus modifying their behaviour (Foucault 1977), Amazon's system of control ensures that every movement is not only actually being observed, but that those observations are recorded, compiled within a database, and algorithmically applied to raise productivity targets or enact sanctions, a system, which closely approximates the Deleuzian notion of constant modulation and computerised control. Katherine Hayles (2012) has recently argued that having surrounded ourselves with intelligent machines, we are becoming inclined to treat people as objects, and perhaps the Amazon warehouse system exemplifies this.

Digital Waste

The discarded remnants of digital architecture form one stream of contemporary electrical and electronics waste (commonly termed e-waste), the fastest growing global waste stream, whose volume was expected to exceed 50 million tonnes in 2015 (Huisman 2012). E-waste is classified as toxic waste, due to the presence of hazardous materials such as lead, polyvinyl chlorides, and antimony in microelectronics. Consequently, for most OECD nations (with the USA being a notable exception), it is illegal to export e-waste to non-OECD nations under the Basel Convention, an international framework which sought to stop the flows of toxic waste from the global rich to the global poor. Additional forms of regional e-waste legislation, such as the EU's Waste Electrical and Electronics Equipment directive complement the Basel Convention, mandating that European e-waste must be recycled at local, high-tech facilities where worker and environmental health and safety regulations ensure that the bodily harms perpetuated within manual e-waste reclamation practices are avoided. Despite these legislative frameworks, there exists a burgeoning illegal international trade in e-waste, which primarily involves shipping material deliberately mislabelled as working second-hand goods for resale in emerging markets. This illegal trade exists because dumping toxic waste upon impoverished people is "cheaper" than sending it for high-tech recycling (Interpol 2009). Cost here is singularly reduced to its economic dimension, the ecological and social costs of e-waste export are externalities whose lack of presence in financial terms permits the perpetuation of a system which enacts serious damage to human and nonhuman bodies and systems, albeit those existing within spaces far from the space of Western consumption flows.

In regions of China, Nigeria, Ghana, and Pakistan, artisanal workers manually "recycle" digital devices. Recycling here is far removed from the vernacular connotations of ecologically responsible and sustainable action, with the practices employed within manual e-waste processing involving the reclamation of specific components and materials which have re-sale value from digital detritus using hazardous methods which detrimentally impact upon workers, local communities, and ecosystems. For example, plastic casings are burned off wiring in order to retrieve the valuable copper wires despite the presence of brominated flame retardants within the casings, which release dioxins and furans, organically persistent substances known to damage nervous and immune systems and cause cancers and harm to numerous internal organs within living systems (Wang et al. 2009, p. 790). The workforce conducting this labour is primarily composed of informal workers, including large numbers of children, who receive financial recompense of around US$1.50 per day for undertaking this hazardous labour. Non-Governmental Organisation (NGO)-led interviews reveal that informal e-waste workers typically lack a formal education, and are unaware of the damage they are doing to both themselves and the local environment by undertaking this form of labour (Roman and Puckett 2002, p. 26). One example illustrating this arises from Guiyu, China, one of the world's largest e-waste processing centres, with an estimated 100,000 artisanal e-waste workers. NGOs exploring the impacts of the local e-waste industry found the local river contained lead levels over 190 times above the limit prescribed by the World Health Organisation (BAN and SVTC 2002, p. 22), and subsequent research revealed that local children had elevated lead levels within their blood (Guo et al. 2014). Lead poisoning is particularly debilitating amongst children due to its detrimental effect upon the development of nervous and neurological systems.

E-waste demonstrates that the global flows of materials surrounding microelectronics do not just move towards the centres of the space of flows, they often return to the margins, causing further damage to impoverished people. Digital technoculture cannot be understood as simply being characterised by a division between the haves and have-nots, as is typically intimated by the discourse of the digital divide, as the life cycle of microelectronics involves individuals, communities, and assemblages of nonhumans who would conventionally be termed digital have-nots, who, as we have seen, are often detrimentally impacted by the ecology of practices surrounding digital technologies.

Conclusion

Examining the life-cycle of microelectronics, we observe that throughout the process—from the extraction of the raw materials through to the processing of e-waste—there are practices which produce deleterious results for the bodies of the human labourers involved and which, in many cases, have additional detrimental effects upon humans and nonhumans not directly associated with the microelectronics industry. These effects occur within a variety of spaces, located in places which traverse the distinctions placed between developing and developed states, but primarily impacting upon workers whose labour is precarious, dispensable, and above all, cheap. Effectively, these invisible bodies are disposable cogs within the cybernetic assemblage of digital technoculture.

As Richard Maxwell and Toby Miller (2012) illustrate, using toxic materials or employing such inequitable labour practices in technological production is not unique to the 21st century and microelectronics. A century ago, a significant proportion of workers involved with the production of batteries in America suffered lead poisoning (Penrose 2003, p. 3) paralleling the e-waste workers and residents of Guiyu, whose lives are similarly blighted by lead poisoning. Impoverished workers and nonhumans suffering from industrial production is a dynamic which long predates digital technoculture and relates to the tendency for profit-orientated enterprise to generate negative externalities whose costs are borne by social and ecological systems. Reducing the concept of cost to a quantitative financial value produces an imperative for industry to displace other costs—in terms of the health and wellbeing of workers, communities, and ecosystems—onto actors who lack the communicative and political capacity to contest inequitable practices.

It should be noted that the singular focus on financial gain appears to be identity neutral, but the manner by which financial power privileges pre-existing economic elites frequently works to reinforce existing inequalities surrounding race and gender. In the examples explored here, the flows of harm towards impoverished humans in regions within China and Nigeria occur within a logic of financial globalisation, which sees corporations move towards locations where wages are lowest and legislation protecting workers and local ecosystems are weakest. However, the pre-existing dynamics of global wealth in a postcolonial world dictate that this predominantly occurs far from New York, London, and Paris. At the same time, China forms a larger market for consumer microelectronics than many former colonial powers in the current techno-capitalist environment. Instead of discrete boundaries between world powers (as in an "East and West" binary), what we find is that within the space of flows there are huge disparities within nations and regions, and that despite neoliberalism's appearance of a race and gender neutral marketplace, historical inequities maintain a presence, albeit not through formal striation. Arguably, this exemplifies the way that the networked logic of neoliberalism turns boundaries into thresholds (Hardt and Negri 2004, p. 55) in the context of the life-cycles of consumer electronics.

What does appear unique within the present system is the globalised space and condensed time-frame in which this life-cycle unfolds. Materials often traverse the globe multiple times during life-cycles of a few years, a process which requires a vast apparatus of transportational infrastructure and the availability of immense quantities of fossil fuels. Additionally, the modes of production and of biopolitical control exercised over the various stages of the process are highly heterogeneous, and problematise any attempts to place such activity within a universal framework of production. The largely unregulated and unmanaged activities of artisanal miners and e-waste salvage workers gesture towards a pre-Fordist labour model, which contrasts strikingly to the organised and disciplinary Fordist regimes present within manufacturing complexes such as Foxconn's factories or the Baotou REE mine. The Amazon warehouse and it's chaotic storage and GPS monitored workers presents another approach, utilising pervasive media technologies to quantify physical action with an unparalleled level of precision, denoting an intensification of previous regimes of biopower and disciplinarity, and correlating with elements of the Deleuzian model of modulating, cybernetic control. This heterogeneous array of production and disciplinary models could be characterised as a postmodern hybrid, but is perhaps better understood as a way of foregrounding the pragmatics of neoliberal capitalism, which is able to adapt to specific local legislative and economic conditions and integrate these modes into a singular drive towards financial efficiency. However, as we have seen, this cybernetic drive towards an efficient system is one which consistently externalises harms upon humans and nonhumans across spaces and modes of production.

The question, then, turns to what can be done to address these highly inequitable situations. Whilst Jodi Dean (2009) makes a compelling case that communicative capitalism is characterised by a fundamental breakdown between public dialogue and political action, it is equally true that without harnessing the affordances of the attention economy and raising awareness of these issues there will not be widespread demand for collective action redressing these issues. Foregrounding the cyclical interplay of production, consumption, and waste decentres the focus upon digital technologies as technological black boxes, short-lived objects of desire within a culture of rapid technological obsolescence, instead highlighting their existence within and across a range of material processes. Such communicative action, and many of the nascent attempts at addressing the issues outlined here, are themselves dependent on digital culture, thus denoting that such an argument does not demarcate technologies as "bad". Rather, it means considering how specific political interventions and mobilisations can bring about positive changes for the often invisible human and nonhuman bodies which exist outside of the privileged spaces of communicative capitalism, and are consequently left out of current conversations about digital media and materiality.

Note

1. See Catilin Overington and Thao Phan this volume for a discussion of the banality of drone technologies and their seemingly impalpable incorporation into everyday forms of urban life.

References

Agence France-Presse. (2014). Apple under fire again for working conditions at Chinese factories. *The Guardian*, 19 December. http://www.theguardian.com/technology/2014/dec/19/apple-under-fire-again-for-working-conditions-at-chinese-factories. Accessed 27 Aug 2015.

BAN and SVTC. (2002). *Exporting harm, the high-tech trashing of Asia.* http://www.ban.org/E-waste/technotrashfinalcomp.pdf. Accessed 27 Aug 2015.

Barboza, D. (2009). IPhone maker in China is under fire after a suicide. *The New York Times*, 26 July. http://www.nytimes.com/2009/07/27/technology/companies/27apple.html. Accessed 27 Aug 2015.

Barboza, D. (2011). Workers sickened at Apple supplier in China. *The New York Times*, 22 February. http://www.nytimes.com/2011/02/23/technology/23apple.html. Accessed 27 Aug 2015.

Bilton, R. (2013). *Amazon: The truth behind the click.* Panorama, BBC 1, 25 November.

Bilton, R. (2014). *Apple's broken promises.* Panorama, BBC 1, 18 December.

Braidotti, R. (2013). *The posthuman.* Cambridge: Polity.

Branigan, T. (2010). Chinese workers link sickness to N-Hexane and Apple iPhone screens. *The Guardian*, 7 May. http://www.theguardian.com/world/2010/may/07/chinese-workers-sickness-hexane-apple-iphone. Accessed 27 Aug 2015.

Castells, M. (1996). *The rise of the network society: The information age: Economy, society and culture volume I.* Massachusetts: Blackwell.

Chen, B. (2010). Workers plan to sue iPhone contractor over poisioning. *Wired magazine*, 14 May. http://www.wired.com/2010/05/wintek-employees-sue/. Accessed 27 Aug 2015.

Coole, D., & Frost, S. (Eds.). (2010). *New materialisms: Ontology, agency, and politics.* London/Durham: Duke University Press.

Cubitt, S. (2014). Decolonising ecomedia. *Cultural Politics, 10*(3), 275–286.

Damasio, A. (2008). *Descartes' error: Emotion, reason and the human brain.* New York: Random House.

Dean, J. (2009). *Democracy and other neoliberal fantasies: Communicative capitalism and left politics.* Durham: Duke University Press.

Deleuze, G. (1992). Postscript on the societies of control. *October, 59*(3), 3–7.

Electronics Take-Back Coalition. (2011). Facts and figures on e-waste and recycling. 25 June. http://www.electronicstakeback.com/wp-content/uploads/Facts_and_Figures_on_EWaste_and_Recycling.pdf. Accessed 27 Aug 2015.

Foucault, M. (1977). *Discipline and punish: The birth of the prison* (A. Sheridan, Trans.). New York: Vintage Books.

Fraden, J. (2010). *Handbook of modern sensors: Physics, designs, and applications* (4th ed.). Dordrecht: Springer.

Friedman, M. (1970). The social responsibility of business is to increase its profits. *The New York Times Magazine*, 13 September.

Fuchs, C. (2010). Labor in informational capitalism and on the Internet. *The Information Society, 26*(3), 179–196.

Fuller, M. (2005). *Media ecologies: Materialist energies in art and technoculture.* Cambridge: MIT Press.

Gindrat, A., Chytiris, M., Balerna, M., Rouiller, E. M., & Ghosh, A. (2014). Use-dependent cortical processing from fingertips in touchscreen phone users. *Current Biology, 25*(1), 109–116.

Guo, P., Xu, X., Huang, B., Sun, D., Zhang, J., Chen, X., & Hao, Y. (2014). Blood lead levels and associated factors among children in Guiyu of China: A population-based study. *PLoS ONE, 9*(8), e105470.

Haraway, D. (1991). *Simians, cyborgs and women: The reinvention of nature.* New York: Routledge.

Haraway, D. (2003). *The companion species manifesto: Dogs, people, and significant otherness.* Chicago: Prickly Paradigm Press.

Hardt, M., & Negri, A. (2004). *Multitude: War and democracy in the age of empire.* London: Penguin.

Haxel, G. B., Hedric, J. B., & Orris, G. J. (2005). Rare earth elements: Critical resources for high technology. *US Geological Survey*, 17 May. http://pubs.usgs.gov/fs/2002/fs087-02/. Accessed 27 Aug 2015.

Hayles, N. K. (1999). *How we became posthuman: Virtual bodies in cybernetics, literature, and informatics.* Chicago: University of Chicago Press.

Hayles, N. K. (2012). *How we think: Digital media and contemporary technogenesis.* Chicago: University of Chicago Press.

Head, S. (2014). *Mindless: Why smarter machines are making dumber humans.* New York: Basic Books.

Huisman, J. (2012). Eco-efficiency evaluation of WEEE take-back systems. In V. Goodship & A. Stevels (Eds.), *Waste Electrical and Electronic Equipment (WEEE) handbook*. Oxford: Woodhead Publishing.

Humphries, M. (2013). Rare earth elements: The global supply chain. *Congressional Reseach Service*, 16 December. http://fas.org/sgp/crs/natsec/R41347.pdf. Accessed 27 Aug 2015.

IEEE. (2013). IEEE experts identify the fourth R-word in sustainability: Repair. *IEEE*, 22 April. http://www.ieee.org/about/news/2013/22april_2013.html. Accessed 28 Aug 2015.

Interpol. (2009). Electronic waste and organised crime: Assessing the links. *Trends in Organized Crime, 12*(3), 352–378.

Ives, M. (2013). Boom in mining rare earths poses mounting toxic risks. *Yale Environment 360*, 28 January. http://e360.yale.edu/feature/boom_in_mining_rare_earths_poses_mounting_toxic_risks/2614/. Accessed 27 Aug 2015.

Joseph, C. (2006). The stark reality of iPod's Chinese factories. *Mail on Sunday*, 11 June. http://www.dailymail.co.uk/news/article-401234/The-stark-reality-iPods-Chinese-factories.html. Accessed 27 Aug 2015.

Kaiman, J. (2014). Rare earth mining in China: The bleak social and environmental costs. *The Guardian*, 20 March. http://www.theguardian.com/sustainable-business/rare-earth-mining-china-social-environmental-costs. Accessed 27 Aug 2015.

Kelly, K. (1994). *Out of control: The new biology of machines, social systems and the economic world*. Reading: Addison-Wesley.

Kitchin, R., & Dodge, M. (2011). *Code/space: Software and everyday life*. Cambridge, MA: MIT Press.

Lazzarato, M. (1996). Immaterial labor. In P. Virno & M. Hardt (Eds.), *Radical thought in Italy: A potential politics*. Minneapolis: University of Minnesota Press.

Liu, C. H., Huang, C. Y., & Huang, C. C. (2012). Occupational neurotoxic diseases in Taiwan. *Safety and Health at Work, 3*(4), 257–267.

Malabou, C. (2008). *What should we do with our brain?* New York: Fordham University Press.

Maxwell, R., & Miller, T. (2012). *Greening the media*. Oxford: Oxford University Press.

NASA. (2012). Rare earth in Bayan Obo. *NASA Earth Observatory*, 21 April. http://earthobservatory.nasa.gov/IOTD/view.php?id=77723. Accessed 27 Aug 2015.

Parikka, J. (2011). FCJ-116 media ecologies and imaginary media: Transversal expansions, contractions, and foldings. *The Fibreculture Journal, 17*, 34–50.

Parikka, J. (2012). New materialism as media theory: Medianatures and dirty matter. *Communication and Critical/Cultural Studies, 9*(1), 95–100.

Parikka, J. (2015). *A geology of media*. Minneapolis: University of Minnesota Press.

Pascual-Leone, A., & Torres, F. (1993). Plasticity of the sensorimotor cortex representation of the reading finger in Braille readers. *Brain: A Journal of Neurology, 116*(1), 39–52.

Penrose, B. (2003). Occupational lead poisoning in battery workers: The failure to apply the precautionary principle. *Labour History, 84*, 1–19.

Pugliese, J. (2007). Biometrics, infrastructural whiteness and the racialised zero degree of nonrepresentation. *boundary 2, 34*(2), 105–133.

Roman, L. S., & Puckett, J. (2002). E-scrap exportation: Challenges and considerations. In *Electronics and the Environment, IEEE International Symposium* (pp. 79–84). doi: 10.1109/ISEE.2002.10032343.

Rose, N. S., & Abi-Rached, J. M. (2013). *Neuro: The new brain sciences and the management of the mind*. Princeton: Princeton University Press.

Schulze, R. (2014). Europium use in TVs and lamps. *EREAN*, 14 April. http://erean.eu/wordpress/europium-eu-used-in-tvs-and-lamps/. Accessed 27 Aug 2015.

Sharma, P., Kaushal, B., & Jain, S. (2013). Study of single and multi wavelength (WDM) EDFA gain control methods. *International Journal of Engineering Trends and Technology (IJETT), 4*(5), 1424–1427.

Sterling, B. (2013). Spime watch: Chaotic storage. *Wired Magazine*, 6 August. http://www.wired.com/2013/08/spime-watch-chaotic-storage/. Accessed 27 Aug 2015.

Stiegler, B. (1998). *Technics and time 1: The fault of epimetheus*. Stanford: Stanford University Press.

Taffel, S. (2013). Scalar entanglement in digital media ecologies. *Necsus European Journal of Media Studies, 3*(Spring). http://www.necsus-ejms.org/scalar-entanglement-in-digital-media-ecologies/. Accessed 27 Aug 2015.

Terranova, T. (2012). Attention, economy and the brain. *Culture Machine, 13*, 1–19.

Thrift, N. (2008). *Non-representational theory: Space, politics, affect*. London: Routledge.

Wang, H. M., Yu, Y. J., Han, M., Yang, S. W., li, Q., & Yang, Y. (2009). Estimated PBDE and PBB congeners in soil from an electronics waste disposal site. *Bulletin of Environmental Contamination and Toxicology, 83*(6), 789–793.

Weiser, M. (1991). The computer for the 21st century. *Scientific American, 265*(3), 94–104.

Zielinski, S. (2006). *Deep time of the media: Toward an archaeology of hearing and seeing by technical means*. Cambridge, MA: MIT Press.

Part III

Biopolitics

This book began with a discussion of the management of bodies and populations through technologies of security and health. Contributors throughout the collection have drawn from Foucault's theories of bio-power (1978) to account for the global inequalities in the cultivation of healthy bodies, and those "deserving" of life, in some spaces and not others. The three chapters which make up the final section of this collection culminate in a sustained focus on biopower and the designation of "unproductive" and "failed" bodies.

Between the two poles of disciplinary anatomo-politics and biopolitics, biopower as a concept accounts well for the myriad systems of training, direction, fluctuation, and normalisation that occupy and structure contemporary Western societies. Elaborated in *Discipline and Punish* (1977), the anatomo-politics of discipline refer to the institutions, such as education and medicine, which pre-conditioned and matured alongside capitalism, focused on training individual bodies in docility and productivity. Correspondingly, in *The History of Sexuality* (1978), Foucault described the biopolitics of population as the grip on and modulation of abstract processes and trends that become salient only at the level of the mass. The emergence of biopolitics accompanied the first recognition of the population as a manipulable object—of 'man as species'—and it thus involved 'having a hold on things that seemed far removed from the population, but which, through calculation, analysis, and reflection, one knows can really have an effect on it' (Foucault 2009, p. 100). Whereas 'sovereign' power before it was the power to administer death, biopower is power over life, the power to 'make live and let die' (Foucault 2003).

The essays that follow examine how three different biopolitical technologies make different groups live, or threaten to institute new controls in the euphemistic name of "security". Caitlin Overington and Thao Phan begin the section by asking what implications the transition of drone technology from military to urban spaces might have for biopolitical governance. David-Jack Fletcher analyses nursing homes in the context of biopolitics and governmentality, as a complex site of quarantine for aged bodies. In the final chapter, Jillian Kramer examines Australia's 'Northern Territory Intervention' into Aboriginal communities as an explicitly biopolitical effort by a self-preserving settler state, with the figure of the 'house' as a site for contestation and resistance.

Biopolitics must be understood as a multifarious strategy of power. Although these chapters deal with three very different processes and technologies, the essential objective of a compartmentalised, securitised, normalised population can be discerned in each of them. Drones in civilian contexts are analysed in terms of banality and Empire (Hardt and Negri 2005): the unproblematic uptake of military tools in non-military space reflects the biopolitical strategy to maintain a generalised, exceptional state of war, wherein heightened surveillance, state violence, and discrimination are justified. A population's health, for biopolitics, depends on the ability of its constituents to contribute to the functioning of capitalism. Aged bodies, from this perspective, can only sap the population's strength. The relegation of the elderly to sites removed from broader society thus reflects biopolitical objectives, while the potential for inhabitants to desire their own removal marks out the nursing home as a junction of overlapping and complex strategies. Finally, the Australian settler state appeals to the security of the (white, colonial) Australian subjectivity, by discursively producing the Aboriginal Other as a threat that must be "stabilised". Biopolitics draws on evolutionist logic to position racial difference as a danger to the population's purity, instituting 'caesurae' that delimit which forms of life can live and which must be destroyed. Little other contemporary legislation operates this logic so brazenly.

References

Foucault, M. (1977). *Discipline and punish: The birth of the prison* (A. Sheridan, Trans.). Harmondsworth: Penguin Books.

Foucault, M. (1978). *The history of sexuality, volume 1: An introduction* (R. Hurley, Trans.). New York: Pantheon Books.

Foucault, M. (2003). *"Society must be defended": Lectures at the Collège de France, 1975–76* (D. Macey, Trans.). New York: Picador.

Foucault, M. (2009). *Security, territory, population: Lectures at the Collège de France, 1977–78* (G. Burchell, Trans.). Basingstoke: Palgrave Macmillan

Hardt, M., & Negri, A. (2005). *Multitude: War and democracy in the age of empire.* New York: Penguin Books.

8

Domesticating Drone Technologies: Commercialisation, Banalisation, and Reconfiguring 'Ways of Seeing'

Caitlin Overington and Thao Phan

The history of drones is irreducibly a history of military enterprise. The evolution of Uninhabited/Unmanned Aerial Vehicle (UAV) technologies through stages of conflict has seen the drone progress from hobbyist radioplane during the First World War, to battlefield reconnaissance vehicle during the Vietnam War, to fully fledged combat weapon in the present day. Despite their long lineage of military service, it was only in the context of the Global War on Terror and the aftermath of the 9/11 terrorist attacks that UAVs were fitted with missiles for the specific purpose of exterminating designated targets. The deployment of armed drones by the Central Intelligence Agency (CIA) in the hunt for terror suspects resulted in the 2002 covert

The original version of this chapter was revised.
An erratum to this chapter can be found at DOI 10.1057/978-1-137-55408-6_11

C. Overington (✉)
School of Culture and Communication,
University of Melbourne, Melbourne, VIC, Australia

T. Phan
School of Social and Political Sciences,
University of Melbourne, Melbourne, VIC, Australia

© The Author(s) 2016   **147**
H. Randell-Moon, R. Tippet (eds.), *Security, Race, Biopower,*
DOI 10.1057/978-1-137-55408-6_8

killing of six al-Qaeda members travelling by motor vehicle in the Marib Desert, Yemen (Whittaker and Burkeman 2002). This event was the inauguration of drones as high profile assault weapon—beginning a trend that would carry on to future US engagements in Pakistan, Yemen, Somalia, and Afghanistan—and coextensively constituted drones in public discourse as the 'signature weapon of the 21st century' (Boussios 2015, p. 43). Yet curiously, despite this brutal reputation, drones have manoeuvred their way successfully into commercial and domestic markets. Rebranded as the Christmas gift du jour, drone sales on eBay during the 2014 American holiday period averaged at 7,600 drones sold each week (Bi 2015). Gimmicky advertising campaigns featuring drones were also on the rise (Overington and Phan 2016) while Amazon (among others) has been fiercely negotiating the trial testing of its conceptual drone-based delivery system.[1]

This rapid invasion of drones into civilian territory demarcates a significant moment in the history of drone discourse. Now finding applications beyond the literal battlefield, drones are deployed within the analogical battlefield—the urban city. This war is the 'banal war' whose primary biopolitical technique is banalisation in the service of Empire. By Empire here, we refer to Michael Hardt and Antonio Negri's denotation of the term as descriptor for the contemporary global order in which war is a 'general phenomenon, global and interminable' (2005, p. 3). In this configuration, the terms of war are deployed domestically as a technique to maintain social order. The War on Poverty, the War on Drugs, and the War on Terror are all examples in which the rhetoric of war is applied as a strategic political manoeuvre to regulate activity and maintain social hierarchies. Such discourses, Hardt and Negri argue, legitimate 'total mobilisation of social forces for a united purpose that is typical of a war effort' (2005, p. 13). Exercises of power and state violence are sanctioned under the constant threat of imminent danger, as 'defence' (the protection against external threats) is reconfigured as 'security' (a constant state of martial activity in the homeland and abroad). Subsequently, the distinction between territories of conflict and territories of peace are blurred and, insofar as war becomes a regular procedure and a means to control social life, war constitutes a regime of biopower within the urban city.

It is our contention that the civilian uptake of drones is demonstrative of the banalisation of war within this regime. Banalisation functions as biopolitical technique to further normalise the protocols of war, to legitimise them as not only practical but necessary within a secure society. Through a comparative analysis with Closed Circuit Television (CCTV), we will discuss the banalisation of security as a general phenomenon and also examine the tensions in the decentralisation of security technologies. Further, the configuration of drones as mediating particular 'ways of seeing', which reproduce surveillant paranoias at the level of the body are, in this analysis, coextensive with Hardt and Negri's Empire. In the same way that war as biopolitical regime denies the distinction between territories of conflict and territories of peace, drones deny a distinction between public and private space. Efforts to rebrand and reconfigure drones as banal entertainment are here critiqued as part of the disciplinary efforts within control societies to perpetuate the discourse of global war.

Drone Terminologies

Although today they are officially referred to in military discourse as Uninhabited/Unmanned Aerial Vehicles (UAVs) or Uninhabited/Unmanned Combat Aerial Vehicles (UCAV), a quick probe into the continued popular use of the term 'drone' also provides a convenient insight into the history and evolution of drone technology. The term 'drone' is often construed as a reference to the low humming sound produced by most early aircraft; however in this instance, it is taken as a reference to male stinger-less honeybees whose primary role is to mate with a fertile queen. These 'drones' do not contribute to hive life and are often dispensed with after mating. In its earliest applications, drone aircraft were primarily used as radio-controlled targets for training antiaircraft gunners (Zaloga 2012). Like drone bees, they were unarmed, expendable, and usually controlled remotely from another, larger aircraft nearby. The term is also a homage to the earliest models of target drones produced by the United Kingdom's Royal Navy, specifically the Kettering Bug and the DH82B Queen Bee—the most commonly used target drone from 1934 to 1943 (Zaloga 2012, p. 7). Later reconnaissance models developed during the Cold War seemed to also reference these entomological roots with names such as Firebee and Lightning Bug drones. Lightning Bugs particularly dominated during the Vietnam War, with the US deploying over 1000 of these drones to fly a total of over 3,400 sorties over China, North Vietnam, and North Korea (Zaloga 2012, p. 15). This was the first ever large-scale use of drones in combat and interest in their surveillant capacities quickly spread internationally.

The current terms UAV and UCAV emerged in the 1980s and are often used to describe the more technologically sophisticated endurance models, such as the RQ-4A Global Hawk (the US military's largest spy drone, capable of intercontinental flight and endurance over two days without refuelling) and the RQ-1 Predator and MQ-9 Reaper (hunter-killer drones equipped with both remote and autonomous piloting capabilities). Whereas the name Global Hawk is a neat juxtaposition to the term drone signifying an evolution from tiny insect to bird of prey, the names Predator and Reaper are alarming departures that bluntly communicate their status as harbingers of death.

Civilian drones belong to a separate taxonomy altogether. Originally developed for military applications, these drones are classified as miniature UAVs and are generally designed to be 'man-portable', that is, small enough to be carried by infantry personnel. Like their larger counterparts, miniature UAVs are usually divided into two categories—those that are used for reconnaissance and surveillance purposes, and those that are designed to carry payloads (Boussios 2015, p. 43). Although the traditional wing and tail models dominate in both categories, the more recent quadcopter models are finding increasing utility in the reconnaissance and surveillance category. The advantage of quadcopters is that they have vertical take-off and landing capacity (VTOL) meaning that they require no launch equipment, can be mounted with gimbals to carry high-precision cameras, can hover in a fixed position, and are often controlled through a tablet interface rather than joystick control, which means they require minimal training to operate. It is for these reasons that quadcopter drones have found wide appeal outside of the military and have smoothly entered the civilian market.

In 2014, global expenditure on drones was estimated at US$6.4 billion with annual expenditure expected to double within the next decade (Bi 2015). Although civil expenditure accounts for a relatively small portion (11 per cent), this number was still great enough to generate a total of $16.6 million in civil drone sales between March 2014 and January 2015 on eBay alone (Bi 2015). Sales were given an extra push during the holiday period with the growing popularity of drones as Christmas gifts officially launching civil UAVs into the mainstream market. While beginner hobbyist drones can cost as little as $50, serious fliers equipped with Global Positioning System (GPS) guided flight-plans, on-board High Definition (HD) cameras, specialised software packages, and stability systems can cost upwards of $2000. Most mid-to-high range quadcopters cost between $400 and $1000, and like their military counterparts, these drones often have a mix of remote and autonomous piloting systems, camera rig, and the strength to carry various payloads—albeit on a much smaller and less destructive scale.

In this chapter, we will continue to use the term 'drone' in reference to UAVs/UCAVs. This is for two reasons. Firstly, 'drone' continues to be the most popular and identifiable term in public discourse. Less sanitised than the acronyms UAV and UCAV, drone connotes the ominous undertones implicit in phrases such as 'Drone Warfare'[2] and 'Obama's Drone War'. Secondly, drone is the term used colloquially by commercial UAV developers in branding their own products. The conscious choice to borrow directly from military jargon is demonstrable of the banalisation of military terminologies within the context of the urban city. That there is no verbal distinction between these two spheres is a crucial point of entry into our analysis, which argues that as there are no caesurae delimiting military or commercial drones, neither are there caesurae delimiting spaces of war and non-war within the discourse of global war in Empire.

Banality and CCTV

Drone technologies now fulfil a multitude of uses within private commercial contexts. These uses permeate a range of industries, from agriculture to real estate, emergency services, and border control (Popper 2015). Indeed, the very expansion of discussions about the utility of the drone demonstrates a general acceptance of the technology's existence within a high capitalist society. However, the ongoing connection that drone technologies have to security and surveillance warrants further investigation.

Lessening the critical awareness of drones as they move across the sky, regardless of their intention, may absolve the public's attention to drones as they continue to play a dominant role in the Global War on Terror, or indeed as they become a greater enabler of regulating and controlling populations within state borders. Examining the banalisation process of CCTV cameras in city spaces offers one possible insight into the future biopolitical implications of normalised drone technologies. While traditionally it remains fixed within its environment, the purpose of CCTV can still be likened to the purpose(s) of domestic drones in that it also often revolves around the concept of seeing and being seen.

The rapid proliferation of CCTV within city spaces across Western countries—namely the UK, Australia, and the US—has been well documented. While this growth is inextricably linked to local political and cultural contexts (Lyon et al. 2012), that CCTV has rapidly transformed from 'novelty to ubiquity' within these environments is in itself indicative of a broader social acceptance of security technologies (Goold et al. 2013, p. 977). Looking to the UK (where the majority of CCTV studies has been focused), CCTV has become a 'common feature of public life' (Brown 1995, p. 1), but it was not always a presumed figure on or for the street. Assisted by the death of James Bulger (a two year old boy murdered by two teenage boys), CCTV footage of the toddler as he walked hand-in-hand with his killers out of a shopping mall near Liverpool in 1993 produced a socio-political environment wherein the footage did not simply affirm itself as crucial evidence to the case, it also illustrated the need for further CCTVs to be installed in order to prevent future crimes (Coleman and Sim 2000, p. 627). The expansion of publicly owned CCTV networks is thus considerably influenced by public attention to violent crimes, which then legitimates the use of this technology as a public good for preventing crime.[3] As Goold et al. argue, '[f]ew protective devices start life as banal. They at first appear unusual, innovative, exciting or scary' (2013, p. 979). In the case of CCTV, its installation is often preceded by a perception of its transformative ability: to make cities safer and more secure.

While the vociferous approval of CCTV as a 'friendly eye in the sky' sits comfortably within the scope of the risk society[4] (Beck 1992; Wilson and Sutton 2004), its disappearance into the architecture of the city and the general apathy associated with its presence (when it is not actively deterring or resolving crime) is better explained by Goold et al.'s term, the 'banality of security' (2013). CCTV initially attracted attention both in its approval and critique (civil libertarian movements were especially critical of the perceived encroachment on the right to privacy in public spaces); however, it has since transformed into a banal security technology. CCTV in contemporary public discourse is 'rarely subject to attention or concern'; the existence of both public and privately owned CCTV networks are regarded as 'mundane, commonplace, scarcely worthy of comment' so they exist 'largely beyond public discourse of contestation' (p. 978). For Goold et al., CCTV should be described as banal not simply because of its invisibility or taken-for-granted presence in the city, but also because this normality enables 'goods' (commercial objects) to gently condition populations to act 'appropriately' (p. 978). While presently drones attract attention regarding their transformative practices, the debate regarding their impact on the cityscape is already diminishing.

The concept of the 'banality of security' draws inspiration from Hannah Arendt's work on the banality of evil. In her book *Eichmann in Jerusalem*, which documents Arendt's reporting of the trial of Nazi SS officer Adolf Eichmann, Arendt describes Eichmann as 'not Iago and not Macbeth', but rather a man with no motives other than 'extraordinary diligence in looking out for his personal advancement' (1994, p. 287). For Arendt, precisely what was striking about Eichmann in the trial was not a 'diabolical or demonic profundity' but the 'lack of imagination which enabled him' to be part of the genocide of Jewish people during World War Two (p. 287). By not necessarily realising the "evil" of his actions and justifying it as a part of advancing a cause or personal success, Eichmann demonstrated how the normality of actions within particular contexts enables tragic and harmful outcomes, often without critical reflection or attention (p. 287). While an important starting point for understanding how the banalisation of security objects such as drone technologies may enable the normalisation of "evil" practices (or expansion of the battlefield into the city), Goold et al. offer an important extension of Arendt's concept. Firstly, they argue that the 'banality of security is a double-edged notion' in that banal security objects can serve as a 'basic social good' whilst at the same time potentially 'undermining that security' (2013, p. 993).

Part of feeling secure is the 'taken-for-granted confidence in the human and non-human infrastructure' surrounding citizens, indicating that not all banality is entirely evil, but rather something that requires vigilance to continue the 'quality and reach of democratic governance' (p. 993). Secondly, their focus incorporates the camera itself. As they argue:

> [O]bjects matter—in terms of the ways they both shape relationships and obscure the exercise of state authority. When objects—particularly security objects—cease to be noticed—these effects can be significantly heightened. (p. 979)

It is this second point through which the potential impact of domestic drones on city spaces can be marked. Drones themselves are not "harbingers of death" but rather enable the expansion of that which is already acceptable. Just as CCTV becomes banal through its attachment and extension of the "beat cop", so too are drones presented as something that enhances already existing desires or practices.

Reflecting the prevalence of CCTV networks within city spaces, most drone technologies and their applications are driven not by public but rather by corporate agencies. Significantly, the number of privately owned and operated CCTV cameras substantially outweighs the number of publicly installed cameras (Reeve 2013). This already has implications regarding the ability to effectively regulate a segregated industry, with many CCTV networks governed loosely by voluntary codes of conduct. The same public/private disparity may be said to exist for the expanding UAV market. While an increasing majority of drones are used in the commercial sector (Popper 2015), the US government also deploys Predator drones to patrol the Mexican border (Associated Press 2014) and US police agencies invest in drones for their policing practices (Pilkington 2014). The influence of the state therefore cannot be dismissed, as evidence of drones enabling population management in some way is already demonstrable. Indeed, the fact that the US Federal Aviation Administration (FAA) has yet to authorise the commercial operation of UAVs for 'non-governmental' purposes—combined with the explosion in companies approved to sell the technology—gestures towards an intimate relationship between state and non-state actors in the drone industry (Federal

Aviation Administration 2015; Popper 2015). Robert Carr's work on the expanding political economy of CCTV illustrates a similarly complex and intimate relationship between state and non-state actors, achieved through the banality of CCTV as a security object (2014). Significantly, as the prevalence and variety of drone applications in cities expands, the threshold of "appropriate" uses will become blurred. With the acceptance of CCTV in city spaces for example, there has been a documented 'shift—or creep—of urban surveillance' (Fussey and Coaffee 2012, p. 201). CCTV now serves a raft of functions in multiple countries, from crime prevention through to improving perceptions of safety; acting as an evidentiary tool through to managing and removing "problematic" populations (Williams and Johnstone 2000; Anderson and McAtamney 2011; Goold 2004; Porter 2009; Norris and Armstrong 1999).[5] Reassembling and expanding its purpose within city spaces, regardless of the effectiveness in its new utility (or the consequences of its new uses), the banality of CCTV enables much of this creep with little sustained critical analysis in public conversation (Goold et al. 2013, p. 983). Indeed, the cliché, "if you've got nothing to hide, you've got nothing to be afraid of", effectively sums up the acceptance of CCTV as being part of the city, along with a demonstration of how difficult it is to meaningfully engage with the impact of security technologies on the basis of the "right to privacy".

CCTV and drone technologies explicitly exist in some form in the city as part of the security and surveillance apparatus of the state. While CCTV continues to illustrate a direct relationship with the state, the proliferation of drone technologies demonstrates a further decentralisation of this surveillance and security assemblage. CCTV, even with its urban security creep, still has some of its primary security functions definitively observed. Exactly what purpose drones serve in the city is harder to articulate, particularly when 'drones designed for use on the battlefield [are] now making their way into less intense civilian applications' for varied purposes beyond strict surveillance or annihilation (Popper 2015). Though their relationship to the state cannot be divorced, it is less distinct than that of CCTV. Drones, as they expand into domestic settings, may not continue as a centralised practice to manage—or destroy—"deviant" and "criminal" groups. Working instead within a broader surveillant and security assemblage, drone technologies within city spaces 'exercise power at multiple sites and through diverse elements that work in conjunction but may also encounter friction' (de Goede 2012, p. 28). In other words, the primary functions of the drone may vary while the net effect remains biopolitical. As in a modern surveillance society, 'power over life and the species' body is not the exclusive attribute of the state, but can be achieved anywhere by any organisation through information gathering and data-management processes and tools' (Ball et al. 2012, p. 38). As drones expand into multiple commercial markets, they continue to collect and store data or enable more efficient movement of capital. As each of these practices becomes more commonplace within the structure of the city, the influence of the technology on these interactions of the everyday becomes less visible, again highlighting the potential consequences of the banalisation of security technologies within cities. Beyond the direct impact of security's banalisation, what becomes of data (which is inevitably created) and how it is used in new contexts or in relation to other information, is less certain. How we interact with the technology though—how it changes our practices in city spaces and how it manages our bodies—is likely to become less visible.

Drones and 'Ways of Seeing'

The way in which the drone moves across city spaces is symbolically significant.[6] However, as many drones in city spaces utilise some form of visual recording, *how* they reproduce city spaces through their own seeing—and how we may be managed by this through the banality of drones in cities—is the focus for this section. Each time we see what a drone sees, we see surveillantly. John Berger's 1972 book *Ways of Seeing* offers a point of entry into understanding the productive impact of the image produced through surveillance, particularly photographs. While photos may at first appear to be a mechanical record that outlasts the object they initially capture, Berger critically argues that every image in fact 'embodies a way of seeing' (p. 10). As Henri Lefebvre infers, 'people look, and take sight, take seeing, for life itself' (1991, p. 75). The image—how it is produced ('we are aware, however slightly, of the photographer selecting that sight from an infinity of other possible sights' [Berger 1972, p. 10]) and how it is consumed—thus represents rather than replicates reality, producing with it expectations of visual order and discourse. That which has been captured is determined to be important, as is the way it has been captured. Represented as something synonymous to or enhancing lived experience, then, 'as consumers we hope to experience that which we never would have been able to without the [drone] technology' (Finn 2012, p. 67). Drones are a technology with a new perspective that allows us to ostensibly see more, to see it all. The perspective that the drone adds something of value within the city, in these perceivably banal aspects of everyday living, continues to inform desires to utilise the technology in new and expanding ways. Recording and storing drone-captured data is ultimately part of what Kevin Haggerty and Richard Ericson (2000) term the 'surveillant assemblage';[7]

> Surveillance is no longer to be conceived as a technology employed by the state in the control of dangerous populations or a tool used by corporations to serve the interests of global capital, but is something that we encounter in advertisements and corporate communication, in video footage for news broadcasts and in our favourite (and least favourite) television programmes and films. Importantly, it is not just something we see—but is something that we do when we post photos and videos to the myriad websites that call for our participation. (Finn 2012, p. 77)

The broad banality of drone technologies, in both their security applications and inherent surveillance functionality, looks set to amplify the problematic de-differentiation between the city space and the war zone, and ongoing critical scrutiny of the drone-produced image is essential. When recording from the sky, drones allow us to see more only by seeing less detail. Their facilitation in more effectively flattening the image of the city from an aerial perspective may therefore be sinister, as the intended functions of a drone do not need to be implicitly related to surveillance or security. Drones will still be part of a biopolitical extension, continually reasserting control through the reproduction and management of city spaces and desire via its securitised ways of seeing.

To reiterate, this technology, as it moves across dynamic city spaces, cannot be bound by one purpose; it will capture all within its frame of reference. Through its expansive purpose of commercial successes, drones become a more prominent 'way of seeing, understanding and engaging with the world' (Doyle et al. 2012, p. 71). The image produced by these cameras becomes authoritative and a representation of truth. Drawing together the discursive power of the image, the assumed benevolence of the banal security apparatus and its legitimation by state and criminal justice processes—approved by the FAA and simultaneously used in police operations for example—these perspectives retain a sense of "premium" representation. Images produced by CCTV cameras are increasingly represented as 'ideal witnesses' in moments of disorder (Evans 2015), and it is not implausible to suggest the same conditions may be met for drones, particularly as they are used in moments of political conflict. By providing new ways of seeing, the practices and populations within cities will also be seen in this "new" light.

Through their flight path, drones seemingly take a step back, to provide the "bigger picture". They are used to capture the spectacular enormity of events and geographies. In doing so, drones reduce the visibility of the micropolitical—of multiple stories—within city spaces, as they can only capture one perspective. Rather than seeing individual bodies in alignment in a protest, drones instead see a mass population, something to be contained or managed. Filming from above, drones— like CCTV cameras—assume a single and authoritative narrative of the space. There is only one "true" story. Through these reproductions, cities are presented as a smooth surface, literally captured by the drone, enclosing it as definable and measurable. As Alison Young contends, expanding on Michel de Certeau's (1984) philosophical account of creative resistance through practices of everyday life in the street, an aerial representation of the city may silence these conflicting narratives: 'the lines of law tend to coincide with the lines of cartography and of timetabling, resulting in the image of the city as smooth, compartmentalised, organised around boundaries and functional, although such a legal assemblage is based on a desire to control the city's perceived unruliness and fecklessness' (2014, p. 42). The desire to see what the drone sees and to participate in the reproduction of these images demonstrates a complicity in the biopolitical desires of the state to control and manage populations within the city.

Drones and the Prosthetics of Empire

The smooth transition of drones from military apparatus to banal object is indicative of broader trends towards the banalisation of security objects within the city. Deployed not only by the state but citizens themselves as part of the rituals of consumption, this kind of quick adoption can be understood as coterminous with what Hardt and Negri (2000) have referred to as 'Empire'. For Hardt and Negri, Empire is representative of a new global order in which traditional notions of sovereignty are complicated by the globalisation of economic, political, social, and cultural exchanges. In particular, the globalisation of capitalist production has altered economic relations so they are increasingly detached from the political control of the nation state. The result is a decline in traditional modes of sovereignty associated with the state, as they write:

> The sovereignty of the nation-state was the cornerstone of the imperialisms that European powers constructed throughout the modern era. By 'Empire,' however, we understand something altogether different from 'imperialism.' The boundaries defined by the modern system of nation-states were fundamental to European colonialism and economic expansion [....] In contrast to imperialism, Empire establishes no territorial center of power and does not rely on fixed boundaries or barriers. It is a *decentered* and *deterritorializing* apparatus of rule that progressively incorporates the entire global realm within its open, expanding frontiers. (2000, p. xii)

Empire is thus symptomatic of a decentralised, deterritorialised 'network power'. But Empire not only describes the reconfiguration of subjectivities at the level of global order, but also the individuation of these subjectivities at the level of the body.

Indeed, Empire functions through the biopolitical production of surveillance techniques as not only normal but also necessary. Taking cues from Foucault, Hardt and Negri have referred to these as the 'productive dimensions of biopower' (2000, p. 27). They argue that Empire is a control society, which produces not only discourse and ideology (a function indebted to the culture industries), but 'agentic subjectivities'—relations, needs, bodies and minds—and embedded within these subjectivities is a language of self-validation and legitimation of authority (pp. 27–33). To aid in this process of legitimation, Hardt and Negri have argued that Empire promotes constant states of global war. War is applied metaphorically by authority, and deployed in domestic as well as foreign contexts. The Global War on Terror is certainly the prime example of this practice—interminable, deterritorialised, substituting a procedural activity for the regulation of bodies within the city. All sites, whether officially classified as 'at conflict' or not, are cast as sites of war insofar as they are structured under the same regime of control. Cities thus become analogic battlefields, with the enemy broadly defined under the label of "terrorist". The banalisation of war is essential in this configuration as war is celebrated as an 'ethical instrument'. Banalisation assists in the biopolitical reproduction of the terms of war as military apparatuses, such as covert surveillance and police repression, are routinised and the enemy is absolutised—cast as 'extremist' and strictly opposed to the existing ethical order (Hardt and Negri 2000, p. 6).

Drones appear in this configuration as part of the iconography of war and, insofar as they are subject to the techniques of banalisation, they too are part of this biopolitical reproduction of global war. In his own analysis of drone technologies in sites of conflict, Joseph Pugliese (2013) has articulated a similar argument for assault drones in service of US military and political interest. In his terms, drones function as part of the 'prosthetics of empire':

> [T]hey extend the imperial power of the state through prosthetic weaponry predicated on violent asymmetries of power. These violent asymmetries of power pivot on an invulnerable/vulnerable axis: while US military personnel can conduct their prostheticised campaigns of militarised violence from the safety of their civil home-sites, the citizens of the countries that are targeted by drone strikes are exposed to a violence that works to obliterate the very difference between civil and military; between civilian and terrorist/soldier. (Pugliese 2013, p. 184)

Like Pugliese, we too see drones as an extension (prosthetic) of interests. But whereas Pugliese's prosthetic is in service of empire (the modernist notion connected to the sovereignty of the nation-state) we see prostheticised drones functioning in service of Empire (the new global order that reproduces subjectivities locked in interminable war). The smooth articulation of one to the other is yet another reflection of the smooth transition of drones across sites that are no longer defined in terms of "war" or "peace".

Conclusion

The invasion of drone technologies into commercial and domestic markets does not demonstrate a fundamental shift in drone functionality; they still continue to work as surveillance and security apparatuses, and are likely to continue to inflict violence, albeit in a more subversive way. This is for a number of reasons. As discussed in this chapter, not only does the appearance of commercial drones within city spaces illustrate how the urban has become the new zone of war in Empire—a space of the ongoing battle for control through "security"—it is likely to enhance these securitisation methods at work within the city. Appropriating discourse and technology, the drones that continue to act as the prosthetics of empire (Pugliese 2013) in distant countries are increasingly present for the Empire within the city. Yet, rather than being perceived as "death from the skies", the same UAVs are celebrated as convenient, entertaining, or enhancing in some way the lived experience of the consumer. Through these applications, drones enable biopolitical impulses of the state to disperse themselves within a greater number of institutions and networks in the city. We are set to become not only accustomed to the presence of drones, but welcoming. By assuming their banality or benevolence rather than engaging critically with drones' transformative agency as they record and reproduce the city through flattened images, the sinister implications of yet more everyday urban surveillance go unchallenged; 'more than a material or technical apparatus—more than a camera—surveillance has become a way of seeing' (Doyle et al. 2012, p. 67). The consequences of this are likely to be profound, as we become complicit in the organisation and management of biopolitically delineated groups. Unlike other banal security technologies within cities, drones are even more decentralised in their applications. Their development and future applications remain to be seen, but domestic drones are likely to be complicit in the biopolitical organisation of populations to enhance neoliberal agendas in capitalist development, accumulation, and domination.

Notes

1. See Sy Taffel this volume for a discussion of Amazon and other corporate uses of surveillance technology to reduce labour, delivery, costs, and time.
2. Often used in news headlines to describe what Joseph Pugliese calls the new paradigm of 'killing-at a distance' (2013, p. 185). See also Pugliese's chapter in this volume.
3. Other prominent examples of CCTV being linked to crime prevention after a crime has occurred include the death of Jill Meagher in 2012 in Melbourne, Australia (Carr 2014), and the September 11, 2001, terror attacks in the United States (Lyon 2004).
4. The risk society, a term developed by Ulrich Beck, describes the way in which modern economic practices have resulted in new and global risks, which are responded to in distinctively modern ways, including a desire to calculate and mitigate (1992).
5. This expansion sits alongside the more generalised emergence of 'pre-crime', a practice relating to the increasingly pre-emptive nature of crime prevention policies—a desire to mitigate and remove potential risks before they become 'real' (McCulloch and Pickering 2009).
6. See, for example, our critique of the way in which drones immobilise different social groups within the #cokedrone ad (Overington and Phan 2016).
7. The term surveillant assemblage builds on Deleuze and Guattari's concept of assemblage, applying it to current understandings, practices, and discourses of surveillance. Haggerty and Ericson's paper is key for expanding discussions surrounding Orwellian or Foucauldian understandings of surveillance, which traditionally focus on surveillance as a centralised practice or specifically targeting "deviant" groups, to explore how various apparatuses work both collectively and individually (and with multiple—sometimes conflicting—objectives) to produce an environment whereby surveillance is a generalised and normalised part of the everyday.

References

Anderson, J., & McAtamney, A. (2011). Considering local context when evaluating a closed circuit television system in public spaces. *Trends & Issues in Crime and Criminal Justice, 430*(October), 1–10.

Arendt, H. (1994). *Eichmann in Jerusalem: A report on the banality of evil.* New York: Penguin.

Associated Press. (2014). Half of US-Mexico border now patrolled only by drone. *The Guardian*, 13 November. http://www.theguardian.com/world/2014/nov/13/half-us-mexico-border-patrolled-drone. Accessed 27 July 2015.

Ball, K., Haggerty, K., & Lyon, D. (Eds.). (2012). *Routledge handbook of surveillance studies.* Oxon: Routledge.

Beck, U. (1992). *Risk society: Towards a new modernity.* London: Sage.

Berger, J. (1972). *Ways of seeing.* London: Penguin.

Bi, F. (2015). Drone sales soar past $16 million on eBay. *Forbes Magazine*, 28 January. http://www.forbes.com/sites/frankbi/2015/01/28/drone-sales-soar-past-16-million-on-ebay/. Accessed 29 Apr 2015.

Boussios, E. (2015). Changing the rules of war: The controversies surrounding the United States' expanded use of drones. *Journal of Terrorism Research, 6*(1), 43–48.

Brown, B. (1995). *CCTV in town centres: Three case studies.* London: Home Office Police Department.

Carr, R. (2014). CCTV installation for crime prevention, or "friends of the government"? *The Saturday Paper*, 2 August. https://www.thesaturdaypaper.com.au/news/politics/2014/08/02/cctv-installation-crime-prevention-or-friends-the-government/1406901600. Accessed 2 Aug 2014.

Coleman, R., & Sim, J. (2000). "You'll never walk alone": CCTV surveillance, order and neo-liberal rule in Liverpool city centre. *British Journal of Sociology, 51*(4), 623–639.

de Certeau, M. (1984). *The practice of everyday life.* Los Angeles: University of California Press.

de Goede, M. (2012). *Speculative security.* Minneapolis: University of Minnesota Press.

Doyle, A., Lippert, R., & Lyon, D. (2012). *Eyes everywhere: The global growth of camera surveillance.* London: Routledge.

Evans, R. (2015). "The footage is decisive": Applying the thinking of Marshall McLuhan to CCTV and police misconduct. *Surveillance & Society, 13*(2), 218–232.

Federal Aviation Administration. (2015). Petitioning for exemption under Section 333. https://www.faa.gov/uas/legislative_programs/section_333/how_to_file_a_petition/. Accessed 12 June 2015.

Finn, J. (2012). Seeing surveillantly: Surveillance as social practice. In A. Doyle, R. Lippert, & D. Lyon (Eds.), *Eyes everywhere: The global growth of camera surveillance*. London: Routledge.

Fussey, P., & Coaffee, J. (2012). Urban spaces of surveillance. In K. Ball, K. Haggerty, & D. Lyon (Eds.), *Routledge handbook of surveillance studies*. Oxon: Routledge.

Goold, B. (2004). *CCTV and policing*. New York: Oxford University Press.

Goold, B., Loader, I., & Thumala, A. (2013). The banality of security: The curious case of surveillance cameras. *British Journal of Criminology, 53*(6), 977–996.

Haggerty, K., & Ericson, R. (2000). The surveillant assemblage. *British Journal of Sociology, 51*(4), 605–622.

Hardt, M., & Negri, A. (2000). *Empire*. London: Harvard University Press.

Hardt, M., & Negri, A. (2005). *Multitude: War and democracy in the age of empire*. New York: Penguin.

Lefebvre, H. (1991). *The production of space*. Cornwall: T.J. Press.

Lyon, D. (2004). Globalizing surveillance: Comparative and sociological perspectives. *International Sociology, 19*(2), 135–144.

Lyon, D., Doyle, A., & Lippert, R. (2012). Introduction. In A. Doyle, R. Lippert, & D. Lyon (Eds.), *Eyes everywhere: The global growth of camera surveillance*. London: Routledge.

McCulloch, J., & Pickering, S. (2009). Pre-crime and counter-terrorism: Imagining future crime in the "war on terror". *British Journal of Criminology, 49*(5), 628–645.

Norris, C., & Armstrong, G. (1999). CCTV and the social structuring of surveillance. *Crime Prevention Studies, 10*(1), 157–178.

Overington, C., & Phan, T. (2016). Happiness from the skies or a new death from above? #cokedrones in the city. Somatechnics, 6(1), 72–88.

Pilkington, E. (2014). "We see ourselves as the vanguard": The police force using drones to fight crime. *The Guardian*, 1 October. http://www.theguardian.com/world/2014/oct/01/drones-police-force-crime-uavs-north-dakota. Accessed 12 June 2015.

Popper, B. (2015). These are the first 500 companies allowed to fly drones over the US. *The Verge*, 7 July. http://www.theverge.com/2015/7/7/8883821/drone-search-engine-faa-approved-commercial-333-exemptions. Accessed 12 June 2015.

Porter, G. (2009). CCTV images as evidence. *Australian Journal of Forensic Sciences, 41*(1), 11–25.

Pugliese, J. (2013). *State violence and the execution of Law: Biopolitcal caesurae of torture, black sites, drones.* London: Routledge.

Reeve, T. (2013). BSIA attempts to clarify question of how many CCTV cameras there are in the UK. *Security News Desk,* 11 July. http://www.securitynews-desk.com/bsia-attempts-to-clarify-question-of-how-many-cctv-cameras-in-the-uk/. Accessed 12 June 2015.

Whittaker, B., & Burkeman, O. (2002). The US accused of executing six alleged al-Qaida members. *The Guardian,* 6 November. http://www.theguardian.com/world/2002/nov/06/usa.alqaida. Accessed 29 Apr 2015.

Williams, K., & Johnstone, C. (2000). The politics of the selective gaze: Closed circuit television and the policing of public space. *Crime, Law & Social Change, 34*(1), 183–210.

Wilson, D., & Sutton, A. (2004). Watched over or over-watched? Open street CCTV in Australia. *Australian and New Zealand Journal of Criminology, 37*(2), 211–230.

Young, A. (2014). *Street art, public city: Law, crime and the urban imagination.* New York: Routledge.

Zaloga, S. (2012). *Unmanned aerial vehicles: Robotic air warfare 1917–2007.* Oxford: Osprey Publishing.

9

The Somatechnics of Desire and the Biopolitics of Ageing

David-Jack Fletcher

Introduction

This chapter examines the ways in which freedom and desire have been discursively constructed as articulations of power through biopolitical and governmental regimes, exemplified in the contemporary biogerontology movement. Adopting a Foucauldian framework of biopolitics,[1] this chapter also examines the construction of ageing as disease in order to mark the connection to somatechnics. Both freedom and desire are conceptualised as highly codified biopolitical systems of laws, rules, and practices that govern the ways in which the body of the individual and the body politic navigate daily life. Freedom and desire are understood here as somatechnic inscriptions, where somatechnics is mobilised as a conceptual framework highlighting the mutual enfleshment and reciprocal relation between soma and techné. That is, between the body

D.-J. Fletcher (✉)
Department of Music, Media, Communication and Cultural Studies,
Macquarie University, Sydney, NSW, Australia

© The Author(s) 2016
H. Randell-Moon, R. Tippet (eds.), *Security, Race, Biopower*,
DOI 10.1057/978-1-137-55408-6_9

and the techniques by which bodies are (trans)-formed (see Pugliese and Giannacopoulos 2009; Sullivan and Murray 2009). Freedom and desire operate within regimes of biopower and, as somatechnic inscriptions, (re)-construct the aged body as not only undesirable, but also as "diseased" within a biomedical framework. Biogerontology, particularly, seeks to eradicate ageing, a notion which this chapter explores through the prominent work of biomedical gerontologist Aubrey de Grey, whose research in this field perpetuates discourses of old age as disease. The disenfranchising and ultimately dehumanising positioning of the elderly in this manner can be examined through a biopolotical lens, which this chapter will deploy in order to account for the development of somatechnologies and regimes of confinement. I argue that biogerontologists operate within these highly structured and codified discourses of desire and freedom in order to not only conduct their research through a biomedical gaze, but also to produce anti-ageing technologies whose goal is the effective "curing" of old age.

Like freedom, both choice and right must also be interpreted as mediated by somatechnologies, for it is through freedom that choices and rights are produced and enacted. Indeed, freedoms and rights are instruments of domination (Foucault 1980) that utilise illusions of individual power while actually subjugating the individual within a constrained set of choices. The notion of choice, then, is also problematic as a manifest form of biopolitical power that is actually constitutive of subjects and individuals. For example, as Michel Foucault explains, 'you can go to prison or join the army, you can go to prison or go to the colonies, you can go to prison or you can join the police' (Foucault 1980, p. 23). The state provides specific choices for the population, however they are structured in such a way that they become less a choice, less a *freedom* of choice, and more a necessity in the sense that the only favourable choice—in those outlined here—is to join a governmentalised regime, for instance the army or the police. Hence, it can be ascertained that *desire* for freedom, as what can be fulfilled through the exercise of a right (to freedom), stems less from an innate human impulse than from the highly regimented articulation of power. My use of the word "desire" is not bound with traditional conceptualisations related to happiness. Rather, desire is understood here as being *produced* through available choices provided by the government and the state. Hence, the desire to be quarantined within a nursing home, as this chapter explores, does not suggest that people necessarily feel they will be happier in these spaces. Rather, these individuals may feel that, given limited options surrounding health care, such spaces are not only their best "choice", but the most practical one as well.

This governmental power is manifold, I argue, and in the context of what I label as the gerontological hygiene movement, the embodiment of this power can be seen in two primary and opposing ways. Specifically, on the one hand, through the production of anti-ageing somatechnologies that perpetuate the notion of old age as disease, and on the other, the construction of, and resettlement of the elderly into, designated medicalised quarantine sites—or compounds—which can be seen as an amended form of Agamben's conceptualisation of camp zones. In this chapter, the work of de Grey is used as an example of the first strategy, the curing of old age, while the Hunter Valley nursing home, in the Hunter Valley region of New South Wales, Australia, is representative of the second strategy—of quarantining the elderly.[2]

The deployment of curing somatechnologies is significant here because the long history of "curing narratives"[3] can be seen to have informed contemporary biomedical practices aimed at extending human labour utility. Indeed, the emergence of modern curing somatechnologies began during the late eighteenth century, almost simultaneously with the deployment of the asylum and other zones of quarantine for undesirable bodies. It is difficult to mark a point of origin for the deployment of curing somatechnologies in relation to ageing, as history is riddled with examples of methods—often cruel by contemporary Western standards—used to preserve youth and cure disease. Contemporary medicalised quarantine sites, such as that of the nursing home, can be understood as compounds, which I will briefly detail later in this chapter, and are connected to somatic concerns of undesirable bodies as somehow unworthy of freedom; in some extreme cases, as unworthy of life.

I argue that desire and freedom, deployed here as somatechnic inscriptions, effectively facilitate, constitute, and regulate bodies at the micro level, allowing choice to operate as a method of both subjectification and subjugation. The individual, then, is produced as some *thing* to be controlled through the very desires, choices, rights and freedoms they believe to be their own. The discourses of freedom and choice, for instance, function to govern not only the movements of specific bodies, but also to orient and channel the desires of social subjects. Situated in this biopolitical regulative framework, the desire for the isolation and quarantine of the insane, the elderly, and the poor can be viewed as a direct result of these discourses of choice, rights, and freedom. Furthermore, these desires can manifest as a sense of obligation, for as Foucault (2009) states,

> in his desire the individual may well be deceived regarding his personal interest, but there is something that does not deceive, which is that the … play of desire will in fact allow the production of an interest, of something favorable for the population. (p. 101)

The desire by many of the elderly to be resettled into medicalised quarantine sites can be seen as exemplary of Foucault's notion above —that is, their removal will be favourable for the population; hence the adoption of a sense of obligation.[4] In this way, these discourses of desire, choice, and freedom have informed historical and contemporary notions of the compound and their loci of control. This is evident in past forms of the asylum, particularly almshouses, which were 'full of broken-down and decrepit men and women … [who were] sent there to die' (Rothman 1971, p. 292). Hence, despite the fact that many elderly people may "choose" to reside in a nursing home, it will be shown here that their "choice" has been shaped by a variety of biomedical, economic, legal, and cultural discourses that position old age as a disease-state, and the elderly as a blight on the social body (see Foucault 1980, 2001, 2012; Cardona 2007; Lombardo 2011; Mykytyn 2010; Rothman 1971). Further, I will go on to argue that this construction of age-as-disease is reminiscent of the humanist quest for immortality.

The Somatechnics of Desire and Freedom: Discourses of Ageing as Disease

Freedom and desire can be seen as powerful tools in this quest for immortality and, as such, are examined here as inscribed by somatechnologies for their ability to reconstitute the body politic and the individual through regimes of biopower. In the domain of biopolitics, freedom is consumed by 'the liberal art of government' (Bröckling et al. 2010, p. 5), which, according to Foucault (2010) 'formulates simply the following: I am going to produce what you need to be free' (p. 63). The production of freedom necessarily produces limitations for what constitutes this freedom and mobilises several regimes of manipulation. For the aged body, this notion of the manipulation of freedom becomes clear when considering medical discourse that overtly states the need for people of a certain age to be resettled into a nursing home. Compounds such as the nursing home can be viewed as an embodiment of contemporary medicalisation practices, where the diseased body may be quarantined for the perceived betterment of more economically productive bodies. As will be shown, these compounds deploy the discourse of freedom to render the medicalised façade of the interiors, which resemble more of a hospital than a "home", curative rather than enclosed.

In a neoliberal context that has seen the reconstitution of bodies and citizens through regulatory and policing practices, the space of the home, wherever this zone is constructed, must be examined as a 'reconfigur[ed] site of surveillance, discipline and forced consent' (Sorrells 2009, p. 1). The medicalised structure of the nursing home enforces particular archetypes of old age that reaffirm the aged body as not only reduced in physical and mental capacity, but also reduced in value as a neoliberal subject. While it can be seen that a hospitalised interior to a nursing home is, to a palliative degree, necessary, many nursing homes deny inhabitants their personal effects and the ability to travel freely to and from the institution, reflecting notions of surveillance and corporeal regulation.[5] In particular, the regulation of movement into—and through—these spaces signifies the mobilisation of gerontological hygiene.

Understanding freedom and desire in the gerontological hygiene movement as inscribed by somatechnologies enables a critical consideration of discourses of disease that seemingly demand corporeal normativity. This demand both produces and is produced by the highly codified systems of laws, rules, and practices that have come to be known as freedom. Intrinsically connected to the somatechnics of desire is the biopolitics of ageing. Discourses that perpetuate corporeal normativity and idealise states of health also produce, necessarily, certain bodies as defective and in need of eradication for the very survival of the human species. The concept of disciplinary anatamo-politics must also be acknowledged here because, while biopolitics operates at the level of the population, anatamo-politics is geared towards the conception of the individual body as a machine. Foucault argues the body is entangled in a nexus of power relations that aim to discipline the body, and to determine both its usefulness and ability to integrate in society, specifically in terms of economic controls (2003b). The individual, then, becomes subject to scrutiny, examination, and (re)-codification.

Utilising both biopolitical and anatamo-political frameworks of health, acceptable bodies and hegemonic norms for the development of somatechnologies to overcome and "cure" diseases—such as old age—can be viewed as a strategy to reactivate humanist notions of acceptable bodies through a re-deployment of eugenic regimes. These biopolitical somatechnologies ultimately seek to perpetuate the humanist quest for perfection and immortality. Where eugenics sought to eradicate 'a life that is already dead because it is marked hereditarily by an original and irremediable deformation' (Esposito 2008, p. 137) through forced sterilisation and liquidation, contemporary biopolitics proposes to eliminate unacceptable "diseased" bodies through corporeal transformation via somatechnologies. While explicitly racist eugenics may have disappeared after the Second World War, I argue that the ideologies and institutional imperatives that initially drove eugenics are ongoing. Indeed, it can be argued that eugenics has never been too far from the minds of many scientists and geneticists, who publicly seek to eradicate various modes of existence and produce ageing as a bodily disease. Although biopolitical methods are subtler than the eugenics practised in America and Europe during the 1900s, they are no less dangerous. The mobilisation of anti-ageing "curing technologies" is beginning to dominate medical discourse with the promise of establishing an eventual cure for the imperfect ageing human. Those who are incurable are often placed within an institution or medical facility, which I position here as a biopolitical regime of quarantine, though the dynamics of this quarantine have altered significantly from historical manifestations, such as the asylum, the Retreat, and the Camp (see Rothman 1971). I argue both instances of curing and having an aged body mobilise a biopolitical biomedical regime. That is, whether you can be "cured" or not, the deployment of anti-ageing technologies and the resettlement of bodies into the nursing home both constitute an attempted erasure of age, thus enacting gerontological hygiene.

In relation to the nursing home, elderly patients in many cases become complicit in both the desire and need for the space of the nursing home for their services, facilities, and access to specialised care. Here the discourse of desire for care is configured, through freedom, as being in the interests of the elderly rather than as a bio-social obligation borne out of corporeal redundancy. While there are no specific laws that mandate the placement of the elderly into these sites, the discourses of the aged as a strain on economy, and of their bodies as visible sites of degradation and inefficiency prove to be quite powerful. These discourses are perpetuated through the emergence of corporations aimed at erasing old age, such as the newly formed Calico and Human Longevity Inc., whose aims are to create and harness advanced technologies in order to manipulate and control the human lifespan. The anti-ageing narratives presented by these companies have been replicated in various scientific palliative care articles. What is strikingly similar in all is the affirmative tone, where the approach to anti-ageing continues to be seen as wholly positive for the elderly and society, ultimately reiterating the view of ageing as problematic. Human Longevity Inc. and Calico espouse holistic approaches to health where the erasure of old age necessarily reduces the risk of acquiring other diseases that occur primarily in bodies with weakened immune systems. The objectives of companies like these can be seen as (in)formed by historical notions of old age as decline. Their research therefore reproduces ageing as problematic, which positions the elderly as a drain on their loved ones and on the social body, incapable of caring for themselves and in need of constant supervision and assistance.

While the elderly are "free" to choose a nursing home (or not), the desire for nursing home assistance has been (re)produced as persuasive through various biopolitical discourses. Desire and freedom are regularly deployed in the way that the gerontological hygiene movement is facilitated. This is achieved partly through the complicity of those elderly subjects who desire their own removal from mainstream society based on the belief that they may indeed be harmful to society. It is important to understand that desire is not understood here as a relation to happiness, rather, following Foucault, as a relation to social duty or obligation, as what needs to be done in order to be a "good" citizen. What I am not suggesting is that the elderly are always "happy" to be removed to these nursing homes. The construction of this desire will be examined presently, in order to trace the emergence of discourses of desire and choice that produce "old age" as a disease-state.

The Gerontological Hygiene Movement: Desiring the Nursing Home

The President's Council on Bioethics report *Beyond Therapy: Biotechnology and the Pursuit of Happiness* (2011) discusses the notion of health and 'wholeness', that is, the human as in possession of both a 'happy soul' and an ageless body. The report continues to discuss the retardation of ageing through 'caloric restriction, genetic manipulations, and prevention of oxidative damage' (p. 173)—research that has been informed by doctors such as de Grey,[6] whose work on the cancellation of ageing has become widely renowned in gerontological communities. What is critical to mark is the reason for this research: why partake in anti-ageing research if not to find a cure for the ageing process? This research, by its very nature, posits ageing as in need of a cure.

Anti-ageing medicine has been described as a paradigm shift in the way humans think about not only disease, but also health. Where 'traditional medicine seeks to treat the complications of ageing, anti-ageing medicine seeks to change the process of ageing in the first place' (California Anti-Ageing cited in Mykytyn 2006, p. 646). Biogerontology has become a contemporary field in which scientists aim to eliminate, or at least pause, the condition of ageing. de Grey, whose research thus far has identified the possibility—indeed, the *probability*—of achieving life extension through technologies, such as his SENS model and emerging telomere and stem cell-based therapies, leaves questions of necessity and desire almost completely unanswered. Removing 'age-related physiological decline' (de Grey 2007, p. 417) is justified as necessary to the evolution of the human, resulting in the erasure of conditions such as disability. His research indicates a resurgence of the millennia old humanist desire for immortality (see Binstock 2003; Juengst et al. 2003). As such, the notion that 'natural death is not the inevitable penalty of life' (Pearl 2009) has been widely mobilised in medical discourse as a justification for life extension research. While the quest for immortality predates notions of ageing as a disease-state, its contemporary manifestation has found its embodiment specifically through a medical framework that understands ageing as a biomedical problem in need of a solution.

This focus on the necessity of anti-ageing technologies and therapies does not negate the influence and significance of desire in the gerontological hygiene movement. The obligation felt by many deemed "old" to be quarantined and isolated acts as an imposition of expectations of age—and ageing—onto their bodies, for 'as individuals, society has a number of culturally and socially defined expectations of how people of certain ages are supposed to behave and how they are positioned and classified' (Powell 2006, p. 8). These classifications of aged bodies may be slowly changing with the development and deployment of a range of anti-ageing somatechnologies, which exist within a scientific discourse of normativity that sees many of the elderly de-valued and disenfranchised as less valuable members of the community. Further, this discourse also positions the elderly as unable to self-regulate and, in extreme cases, as unworthy of life. In this way, then, the biopolitical frame of the right to 'make live and let die' (Foucault 2003b) becomes critical in the quarantine of elderly subjects. After a lifetime of being made to live according to certain rules and expectations posing as freedom, the elderly are relegated to quarantined compounds where they can be 'let [to] die'. More specifically, 'one might say that the ancient right to *take* life or *let* live [has been] replaced by a power to *foster* life or *disallow* it to the point of death' (Foucault 1978, p. 138).

In Western liberal democratic states, the relation between citizens and governmental institutions is founded on the liberal concept of mutual responsibility. The notion of governmentality becomes crucial here, where Foucault's concept is understood to operate through both anatamo- and bio-politics, and encapsulate the way in which 'one conducts the conduct of men' (2010, p. 186). Governmentality explains how self-regulation, for instance obeying state rules and maintaining healthy living, is linked to the broader biopolitical right to make live and let die. Ultimately, it can be seen as a system of power relations between 'institutions, procedures, analyses and reflections, calculations, and tactics' (p. 144) targeted at regulating and controlling the population. This draws into question notions of the self, which Foucault observed as a construction reliant upon multiple forms of knowledge and institutionalised practices, deploying the notion that the self not only exists in relation to multiple forms of power, but is also in fact created by them (Holmes and Gastaldo 2002). It is the connection between the self and institutions that is of focus here, specifically the institution of the nursing home for its relation to technologies of the self and self-regulation.

This notion of self-regulation is evident in relation to the elderly through the mobilisation of medicalised quarantine sites, such as nursing homes, which operate to remove specific individuals from societal view. These spaces can be marked as compounds for their clinical structure and the techniques of surveillance embedded within their walls, which (re)-produce these subjects as problematic through discourses of decline and disease. Arguably, the elderly themselves adopt these discourses, particularly those who desire to resettle themselves in these zones. It is important, however, to acknowledge the marked difference between these sites and historical camp zones (see Agamben 1997), such as those created during the Nazi state. The abhorrent regimes of eradication are not seen in contemporary constructs of camp-like facilities, and to align these with compound structures, such as the nursing home, would not only be unethical, but also highly problematic. The term compound has been used purposefully here in order to account for these zones as both a manifestation of the gerontological hygiene movement and the deployment of desire as mediated by somatechnologies.

Before embarking on an examination of the nursing home as a compound, it must be acknowledged: firstly, that many elderly citizens "choose" to reside in a nursing home—"choose" has been placed under interrogation as this notion will presently be problematised; secondly, these individuals are provided specific medical care tailored towards their lifestyle/s, with the interest of maintaining optimal health for one's age; and thirdly, I do not argue these spaces should not exist, nor that people do not need the care provided. Despite this label of care, critical governmental and biopolitical intrusions remain evident in nursing homes. This is not to suggest, by any means, that these institutions act in the same way as, for example, the refugee camp. However, as Agamben (1997) notes, the camp can be altered for specific purposes and to house specific demographics. The nursing home can be understood as a compound when considering the contemporary hegemonic paradigm, which insists that the elderly have become dependent subjects instantiating a strain on economic, political, and public life (see Cumming and Henry 1961; Phillipson 1998; Powell and Biggs 2003; Powell 2006; Dannefer and Phillipson 2010). It is widely believed that the elderly belong in these spaces and thus must be removed to them, which is reminiscent of the Retreat and the asylum (Rothman 1971). The belief that these individuals need to be removed from "free" society and its attendant freedoms of mobility, essentially disenfranchises them, simultaneously placing them further in the grip of governmental regimes.

Furthermore, isolation and quarantine are pertinent to the nursing home. While this quarantine is neither permanent nor mandatory, the institutionalised values associated with the removal of the elderly into a village space draw on eugenic principles of inferiority. Quarantine may generally be understood as an act of total and complete segregation and isolation, whether temporarily or permanently—however Curtis (2002) mobilises an alternative understanding when he states that 'quarantine operates by identifying and separating out problematic dimensions of social life and social relations and then subjecting them to particular treatment protocols' (p. 514). In this sense the nursing home can be seen as a construct of quarantine for the "betterment" of society. Within this quarantine, security and surveillance are present, purporting to be in the best interest of those inhabiting these zones. While it must be acknowledged that a degree of surveillance is legitimate in order to protect residents from break-ins and other criminal behaviour, certain facilities boast the presence of

> a security camera system installed in all communal areas and corridors enabling us to monitor activities throughout the facility 24 hours a day. A locked keypad system is installed to all entry and exit doors to the building and a contracted security company performs nightly patrols around the grounds. (Hunter Valley Care 2014)

The notion of panoptic surveillance in the Hunter Valley Care facility is implied here as a method of ensuring safety. This security technique is demonstrative of a biopolitical regime of regulation and control. Furthermore, this security technique is implicated in the production of a specific concept of the elderly subject, a subject that is perceived as unable to care for itself. By mentioning this feature of the nursing home, the company no doubt aims to highlight the safety of the residents within; however this does little more than to rearticulate notions of the compound, where those residing within the walls of the nursing home are subject to strategies of compliance through their constant surveillance. Indeed, the residents are unable to come and go as they please, as the facility is locked during the night-time hours. Although there may be provisions in place to allow for late exit/entry from the facility, the notion that these individuals need permission for night-time mobility demonstrates that their freedom is produced through strict limitations that regulate the movements of those within. The compound structure is also disguised by the allowance of personal items, not unlike the decoration of prison cells, in order to 'create the feeling of home' (2014)—though the personal effects are subject to approval and can be removed by the institution without notice. In addition to the surveillance and monitoring of personal effects, Hunter Valley Care offers various levels of service for those who are able to afford it. For instance, under the program entitled 'Extra Services',

> residents … enjoy a higher standard of accommodation, a choice of meals, the option of alcoholic drinks with dinner, and included additional lifestyle activities such as hairdressing and beautician treatments. (2014)

The significance of the above text is the inclusion of 'choice', which implies that those who are unable to afford 'Extra Services' are not entitled to a *choice* of meals, or the *option* of alcohol with dinner and are effectively denied access to certain lifestyle activities. While these options are spruiked as positive features of the Hunter Valley Care facility, they produce discourses of choice and freedom as existing within highly codified rules, practices, and laws, where the level of "care" provided is inextricably bound with the resident's socioeconomic status.

The Hunter Valley Care facility is by no means representative of every nursing home, however the surveillance techniques deployed are standard procedure, as is the clinical compound structure of these spaces. Central to this idea of the nursing home is that of old age and ill health. The nursing home provides this care for those nearing the end of their life path. The notion of gerontology here is quite significant as the criteria as to who is eligible to reside within the space of the nursing home are quite restrictive, and inextricably connected to those who are deemed "old". Indeed, 'old age has simultaneously become a major source of "risk" but also a potential source of "liberation"' (Cardona 2008, p. 480) in that the elderly are "liberated", in Foucauldian terms, from normal civic work and responsibilities and provided a classed form of round the clock healthcare not afforded to other marginalised bodies. Nursing homes may be a site for contestation here, as in many cases people decide it is within their best interest to be resettled in these spaces. However, the paradigm of health surrounding this decision incentivises self-exclusion in exchange for palliative care, and is demonstrative of desire as a discursive construct produced through power relations. The promotion of retirement institutions 'played an important role in bracketing out many fundamental anxieties associated with events, such as the loss of work in early older age, to the loss of bodily function in later life' (p. 480). The implication here is that people have been given a set of choices regarding what sort of care they wish to receive, and they are essentially free to opt for one choice over another. As I have argued, this ability to choose is inevitably connected to one's level of wealth. This classed ability to self-exclude integrates freedom to choose health care into governance of "unhealthy" and "unproductive" bodies and does not erase the functioning of nursing homes within biopolitical frameworks of quarantine and isolation.

The ability to labour, as has been shown, has been connected to one's social value (see Arendt 1998) and the construction of a body as unproductive. The choices made by many of the elderly reflect long-institutionalised discourses that strictly affirm and perpetuate notions of the labour-value dichotomy, and which teach us that the most plausible solution for those bodies with diminished labour capacity is confinement to nursing homes and other compound facilities. These discourses are further justified through the marketing of anti-ageing technologies as somehow necessary, not only for the preservation of youth, but also, critically, for the restoration of an individual's social and economic value. Inherent within anti-ageing technologies, then, is the desire to keep bodies as productive as possible for as long as possible, thereby rendering the medicalised quarantine site of the nursing home as a logical solution when all else fails.

Conclusion

Framing desire and freedom as mediated by somatechnologies enables us to examine the various practices, laws, and regulations that govern the bodies of both the individual and the population. By producing, effecting, and channelling our desires, the apparatuses of governmental power are able to thereby regulate the ways in which we use our bodies, view our bodies, and treat the body of the Other—in this case, the elderly. This governmental framework reveals how old age has been conceptualised as a disease-state through a biomedical lens. The work of biogerontologists such as de Grey (2005, 2007) demonstrates the powerful humanist discourses of immortality that are prevalent in contemporary society and biomedicine. Critically, de Grey's work is not questioned in terms of necessity; rather, his work is seen as critical biomedical research in the scientific community, and his conceptualisation of ageing as disease-state is regularly accepted as a common-sense fact. In what I have framed as a gerontological hygiene movement, the elderly are subjugated by regimes of resettlement into medicalised quarantine sites and through the somatechnic inscription of desire, are encouraged to believe that this resettlement is not only best for them, but also for society at large.

The biomedical gaze has been central to this chapter's tracing of the emergence of biopolitical discourses of disease and practices of eradication through the deployment of highly regimented discourses of desire and freedom that produce the elderly body as diseased. Drawing upon Foucauldian biopolitics, this chapter has critically addressed the formation of both desire and freedom, which enable—indeed, justify—the resettlement of the elderly into the compound space of the nursing home, where several of the humanist ideologies concerning acceptable bodies and immortality remain. Historically, the notion of desire has been codified through a series of rules, practices, and laws that provide the population with the illusion of freedom. This illusion has perpetuated humanist ideologies of immortality as seen through the medicalisation of old age and the consequent "desire" for the elderly—including, *by* the elderly—to be resettled into compound spaces.

This chapter has focused on the medicalisation of old age and the discursive production of the gerontological hygiene movement; however, it is crucial to mark the trajectory of this movement with the proliferation of anti-ageing somatechnologies that seek to not only halt the ageing process but to reverse it entirely. This movement is already progressing beyond the resettlement of the elderly into medicalised quarantine sites, it has now shifted towards hard technologies—also somatechnic in nature—that permeate the boundaries of the flesh in an attempt to violently reclaim youth. The implication here is that regimes of hygiene are far from historical tragedies; they are what will potentially haunt the future.

Notes

1. Biopolitics is understood here as a control disposit if for the population, specifically the regulation, manipulation, and conditions of birth, life (and life expectancy), and death (Foucault 2010).
2. It must be noted that I do not intend to suggest *all* elderly individuals *need* care, nor do I suggest that all elderly individuals desire their own quarantining. More specifically, I frame this desire in terms of an obligation as discussed in this chapter, where, in neoliberal society, the measurement of social use is based on levels of health and productivity.
3. Foucault (2003a) maps the development of these technologies in the late eighteenth century, arguing their emergence was concurrent with the discursive shift in medicine and understandings of the body.
4. Governmentality of aged care is not linked solely to instrumentalising desire for self-exclusion; several external forces, such as the extended working hours of families and the increase of double-income homes,

are also implicated in the removal of aged bodies from the home. For the purposes of this chapter, these other forces are not examined as they are tangential to the primary focus of gerontological hygiene. Further, it must be noted that the notion of desire for quarantine by the elderly is not a universal description of all elderly people in all nursing homes. Rather, it is an extension of Foucauldian theory of biopolitics and governmentality to explain the situation that many elderly people find themselves in.

5. It is important to note that these characteristics are not universal to all nursing homes. However, for the purpose of this chapter, the phrase "nursing home" will denote such characteristics, specifically in the context of the Hunter Valley Care Facility.

6. de Grey is perhaps most well known for his work on caloric restriction and his relatively new SENS model, that is, Strategies for Engineered Negligible Senescence. He has been compiling data on gene therapy, cell therapy, and tissue engineering in order to find a way to combine these therapies to medically treat ageing and, as he puts it, 'break free of the limitations on life span that our imperfect maintenance machinery naturally imposes' (de Grey 2007, p. 417).

References

Agamben, G. (1997). *The camp as the nomos of the modern*. Stanford: University Press.

Arendt, H. (1998). *The human condition: The second edition*. Chicago: University Press.

Binstock, R. H. (2003). The war on "anti-ageing medicine". *The Gerontologist, 43*(1), 4–14.

Bröckling, U., Krasmann, S., & Lemke, T. (2010). *Governmentality: Current issues and future challenges*. New York: Routledge.

Cardona, B. (2007). "Anti-ageing medicine" and the cultural context of ageing in Australia. *Annals of the New York Academy of Sciences, 1114*(1), 216–229.

Cardona, B. (2008). "Healthy ageing" policies and anti-ageing ideologies and practices: On the exercise of responsibility. *Medicine, Health Care and Philosophy, 11*(4), 475–483.

Cumming, E., & Henry, W. E. (1961). *Growing old, the process of disengagement.* New York: Basic Books.

Curtis, B. (2002). Foucault on governmentality and population: The impossible discovery. *The Canadian Journal of Sociology/Cahiers canadiens de sociologie, 27*(4), 505–533.

Dannefer, D., & Phillipson, C. (2010). *The SAGE handbook of social gerontology.* London: SAGE Publications Ltd.

de Grey, A. (2005). A strategy for postponing ageing indefinitely. *Future of Intelligent and Extelligent Health Environment,* 118: 209–219, Amsterdam: IOS Press.

de Grey, A. (2007). The natural biogerontology portfolio. *Annals of the New York Academy of Sciences, 1100*(1), 409–423.

Esposito, R. (2008). *Bíos: Biopolitics and philosophy.* Minneapolis: University of Minnesota Press.

Foucault, M. (1978). *The history of sexuality* (Vol. 1). New York: Pantheon Books.

Foucault, M. (1980). *Power/knowledge: Selected interviews and other writings, 1972–1977.* New York: Pantheon Books.

Foucault, M. (2001). *Madness and civilization: A history of insanity in the age of reason.* Oxon: Routledge.

Foucault, M. (2003a). *The birth of the clinic: An archaeology of medical perception.* London: Routledge.

Foucault, M. (2003b). *"Society must be defended": Lectures at the collége de France, 1975–76.* New York: Picador.

Foucault, M. (2009). *Security, territory, population: Lectures at the collège de France 1977–1978.* New York: Picador.

Foucault, M. (2010). *The birth of biopolitics: Lectures at the collège de France, 1978–1979.* New York: Picador.

Foucault, M. (2012). *Discipline & punish: The birth of the prison.* New York: Pantheon Books.

Holmes, D., & Gastaldo, D. (2002). Nursing as means of governmentality. *Journal of Advanced Nursing, 38*(6), 557–565.

Hunter Valley Care. (2014). Accommodation. *Hunter Valley Care.* http://huntervalleycare.com.au/our-facilities/amaroo/accommodation/. Date Accessed 1 Sept 2015.

Juengst, E. T., Binstock, R. H., Mehlman, M., Post, S. G., & Whitehouse, P. (2003). Biogerontology, "anti-ageing medicine", and the challenges of human enhancement. *Hastings Center Report, 33*(4), 21–30.

Lombardo, P. A. (2011). *A century of eugenics in America: From the Indiana experiment to the human genome era*. Indiana: Indiana University Press.

Mykytyn, C. E. (2006). Anti-ageing medicine: A patient/practitioner movement to redefine ageing. *Social Science & Medicine, 62*(3), 643–653.

Mykytyn, C. E. (2010). Analyzing future predictions: An anthropological view of anti-ageing futures. In G. M. Fahy (Ed.), *The future of ageing: Pathways to human life extension*. New York: Springer.

Pearl, R. (2009). *The biology of death*. UK: Read Books.

Phillipson, C. (1998). *Reconstructing old age: New agendas in social theory and practice*. London: Sage.

Powell, J. L. (2006). *Social theory and ageing*. Maryland: Rowman & Littlefield Publishers.

Powell, J. L., & Biggs, S. (2003). Foucauldian gerontology: A methodology for understanding ageing. *Electronic Journal of Sociology, 7*(2), 1–14.

President's Council on Bioethics. (2011). *Beyond therapy: Biotechnology and the pursuit of happiness*. Minneapolis: University of Minnesota Press.

Pugliese, J., & Giannacopoulos, M. (2009). The lex of somatechnics. *Griffith Law Review, 18*, 207–211.

Rothman, D. J. (1971). *The discovery of the asylum: Social order and disorder in the new republic*. Piscataway: Transaction Publishers.

Sorrells, K. (2009). Bringing it back home: Producing neoliberal subjectivities. *Liminalities: A Journal of Performance Studies, 5*(5), 1–6.

Sullivan, N., & Murray, S. (Eds.). (2009). *Somatechnics: Queering the technologisation of bodies*. Surrey: Ashgate.

10

Securing Sovereignty: Private Property, Indigenous Resistance, and the Rhetoric of Housing

Jillian Kramer

On 14 July 2009, over 100 Alyawarr people left Ampilatwatja, a small Aboriginal town in the centre of Australia. In order to protest the Federal Government's Northern Territory National Emergency Response (known as the Intervention), they moved three kilometres away to communally owned Aboriginal land that had not been annexed by the policy. They wrote to supporters from their new camp at Honeymoon Bore:

> We, elders from the Ampilatwaja community, three hours north-east of Alice Springs, walked out of our houses and set up camp in the bush. We are fed up with the federal Government's Intervention, controls and measures, visions and goals forced onto us from the outside ... The NT Intervention hasn't brought any improvements to our people's lives. It hasn't brought us any new houses ... it's the same old ration days of flour, tea and sugar and some clothing. (Downs 2009b)

J. Kramer (✉)
Department of Media, Music, Communication and Cultural Studies,
Macquarie University, Sydney, NSW,

© The Author(s) 2016
H. Randell-Moon, R. Tippet (eds.), *Security, Race, Biopower*,
DOI 10.1057/978-1-137-55408-6_10

As their reference to the 'old ration days' suggests, the Alyawarr people's walk-off resonates profoundly with the Lake Nash and Gurindji walk-offs of the 1940s and 1960s, respectively (Korff 2015). Their demands were similar. In the face of ongoing dispossession, the Alyawarr people wanted the Government to return their land, end paternalistic and 'protection-ist' forms of regulation, and pay fair wages (Downs 2009b). They also demanded an end to the Intervention.

The Intervention was launched in June 2007. Following the latest in a series of reports about child sexual abuse in Aboriginal commu-nities, former Prime Minister John Howard and Indigenous Affairs Minister Mal Brough declared a 'national emergency' (Howard 2007c). Drenched in militaristic language, they proposed an ostensibly 'radical, comprehensive and highly interventionist' policy that would 'establish law and order,' 'good governance', and 'normality' in seventy-three tar-geted Aboriginal communities (Howard 2007a, b, c). The military and additional police were deployed to 'move in' and 'take control' (Howard 2007b). Against the backdrop of front-page headlines such as 'Martial Law' and 'Crusade to Save Aboriginal Kids,' these forces were charged with implementing a suite of 'special measures' (Adlam and Gartress 2007, p. 1; Karvelas 2007, p. 1). For example, the Government reac-quired communally owned Aboriginal land, imposed alcohol and pornography bans, and ordered that children undergo a medical examination. The policy reproduced punitive measures designed to not only regulate and control the everyday lives of over 45,000 tar-geted Aboriginal people, but also coerce targeted Aboriginal people and their land into the so-called 'real economy' (Brough 2007c, p. 11). The Government developed economic regimes that aimed to obliter-ate communal land holdings, while promoting private home and land ownership and inserting communities into the neoliberal marketplace. Compulsory income quarantining was imposed, the Community Development Employment Program (CDEP) was abolished, and Government Business Managers (GBMs) were sent to seize control of each community's assets, such as earthmovers, community organisa-tions, and supplies.[1] These moves were heralded on the front-page of the nation's only national newspaper, the *Australian*. Accompanied by the byline 'Paternalistic, but plans' time is now,' Nicolas Rothwell wrote in glowing terms: 'the "emergency response" aims at establishing nothing less than a new social order in the bush. This is human engineering on the grand scale' (2007, pp. 1–2).

Protest House

It is precisely these biopolitical attempts at 'human engineering' that the Alyawarr elders rejected. Following their move to Honeymoon Bore, they partnered with several unions to build a new home on communally owned Aboriginal country. It was named 'Protest House.' This compelling site of protest exposes the racial arsenal that I want to explore in this chapter. The Alyawarr people's tactical repurposing of their house as a signifier of their sovereignty over country, law, communal title, and resistance reveals the significance of the "house" as a critical site of contestation. Protest House exposes both the historical and geographic points of connection between Australia's history of white invasion, the ongoing settler-colonial project, and transnational imperial formations. It not only stands as a direct challenge to the Intervention, it also contests the ways in which the figure of the "house" and its attendant political, juridical, and economic infrastructure operate as a colonising apparatus.

Paula Chakravartty and Denise Ferreira da Silva (2012) open up the space to further explore the significance of 'Protest House.' Writing in the context of the sub-prime crisis in the United States that saw disproportionate numbers of Black and Latino/a people's houses repossessed, they identify the racial forces embodied in the figure of the "house." In the opening lines of their essay, 'Accumulation, Dispossession, and Debt: The Racial Logic of Global Capitalism,' they argue:

> A house is a juridical-economic-moral entity that, as *property*, has material (as asset), political (as dominium), and symbolic (as shelter) value. Houses, as such, refer to the three main axes of modern thought: the economic, the juridical, and the ethical, which are, as one would expect, the registers of the modern subject. It is, in fact, impossible to exaggerate the significance of individual (private) property in representations of modernity. (Chakravartty and da Silva 2012, p. 362; original emphasis)

Despite significant historical and contextual differences, Chakravartty and da Silva's (2012) understanding of the "house" resonates in the context of the Intervention into Aboriginal communities such as Ampilatwatja. On one level, they expose the transnational capitalist logics that reproduce racialised subjects as premodern, affectable, and unable to possess a house as they 'lack' the self-determination and property rights 'that distinguish the proper economic subject' (Chakravartty and da Silva 2012, p. 368). On another level, they also elucidate the ways in which the symbolic and material figure of the house is intertwined with representations of private property, sovereignty over country, and modernity.

Following Chakravartty and da Silva's (2012) work, I want to explore the ways in which the rhetoric of housing operates within the Howard Government's plans to reacquire Aboriginal land, enforce 'home owner-ship,' and generate 'private property.' In order to perform this analysis, I want to draw a distinction between the figures of the "home" and the "house." The "home" has been the site of valuable feminist and anti-racist scholarship. For example, the work of scholars such as Sophie Bowlby et al. (1997) and Lorna Fox (2008), perform a feminist critique that exposes the gendered and patriarchal dimensions of the "home" and "home ownership." In her article '"Assimilation Begins in the Home": The State and Aboriginal Women's Work as Mothers in New South Wales, 1900s to 1960s,' Heather Goodall (1995) builds on this analysis. She maps the ways in which assimilation policies operate within Aboriginal subjects' homes in order to surveil, discipline, and assimilate Aboriginal women's bodies and practices. As such, I will depart from these stud-ies of the "home" in order to explore the role of the "house" within the settler-colonial state's economic and juridical structures. This racialised and colonial figure of the house materialises the ways in which the 'ana-lytics of raciality' instantiated with the originary assertion of *terra nullius* work as an *a priori* within colonial law to position white subjects as mod-ern, lawful, and property-bearing subjects (da Silva 2007). The house is consistently mobilised by the settler-colonial state and its subjects in their attempts to assert and legitimate white possession of unceded Indigenous country.

In the work that follows, I argue that the Intervention amplifies the significance of the house within the colonial and racial logics that undergird the settler-colonial state. Before I proceed to make this argument, however, I want to stress that I do not read the Intervention as an "exceptional" or "unprecedented" policy. On the contrary, I argue that this policy represents a continuation of the biopolitical regimes of racial warfare that must be tracked back to white invasion. My use of the word 'biopolitical,' here, draws on Michel Foucault's thesis (2004, p. 257). Specifically, it draws on his understanding of a form of power that developed in the late eighteenth and nineteenth century. Critically intertwined with sovereign power and colonising projects, this form of power targets, hierarchises, and instrumentalises bodies and populations; it implicates bodies and biological processes within complex mechanisms of disciplinary normativity, intervention, and regularisation. Predicated on the position of whiteness as the governing and—as Chakravartty and da Silva (2012) suggest—the "normal" biopolitical category, this form of power fragments the world's population into two groups: those who 'must live' and those who can be 'let to die' (Foucault 2004, p. 254).

'A History That Never Ends': Ongoing Racial Warfare

Rosalie Kunoth-Monks evidences the ways in which the Intervention perpetuates modes of biopolitical warfare. She described the arrival of the Howard Government's forces in Arlparra—or Utopia—a number of small homeland communities in the centre of the Northern Territory just 100 kilometres from Ampilatwatja:

> On the day that soldiers in uniform, the police and public servants arrived and we were ushered up to the basketball stadium and we were all told that we were now under the Intervention. We don't have access to newspapers … a lot of us were going along our normal way … On that day, when they landed, it was incredible. We really thought we were going to be rounded up and taken. (Kunoth-Monks 2012)

Here, Kunoth-Monks' testimony invokes the histories of frontier warfare that have produced lasting intergenerational trauma within targeted Aboriginal communities. Her claim, 'we really thought we were going to be rounded up and taken' gestures towards consecutive policies that have led the police and military to enter Aboriginal communities. Under the 'protectorate system,' for example, the settler-colonial state's forces repeatedly entered targeted Aboriginal communities to "round up" and "take" Aboriginal people from their homelands. The *Northern Territory Aboriginals Ordinance 1911* (Cth) enabled the 'Chief Protector' and police force to lawfully 'enter any premises where the aboriginal or half-caste is or is supposed to be and may take him into his custody' (1911, p. 62). In the 1930s, these powers facilitated a biopolitical pro- gram devised by Chief Protector Dr Cecil Cook to ensure both white dominion and 'the complete disappearance of the black race, and the swift submergence of their progeny in the white' (cited in Markus 1990, p. 93). Compelled by his belief that 'full-blood' Aboriginal people should be left to 'die out' in segregated land 'reserves,' Cook forcibly removed Indigenous children who were considered to have 'mixed blood' from their families. One Indigenous woman described this regime: 'I remem- ber all we children being herded up, like a mob of cattle, and feeling the humiliation of being graded by the colour of our skins for the govern- ment records' (Confidential Submission 332, cited in Australian Human Rights and Equal Opportunity Commission 1997, p. 153). Selected chil- dren were then incarcerated in punitive missions that aimed to eliminate Aboriginal cultures and languages. In a program that continued until the 1960s, they were subject to discriminatory regulations of the Chief Protector until they were assimilated into the white workforce; these chil- dren are now known as the "Stolen Generations."

Against this backdrop of racial terror, Indigenous scholars and activ- ists have rejected Howard's claim that the police and the military must again enter targeted Aboriginal communities in the name of "protecting" Aboriginal children. They have argued that this policy is a continuation of foundational colonial violence. For example, Pat Turner argues that the catch-cry of "protecting children" has been used—just as it was under the 'protectorate system'—as a 'Trojan horse to reassume total control of our land' (cited in Australian Broadcasting Corporation 2007). George Gaymarani Pascoe, an Indigenous elder from Milingimbi, echoes this argument: 'thinking again back in history to the [nineteen] twenties and thirties ... [The] Intervention did literally, I say literally, deliberately ... come across to us as a history that never ends' (cited in Harris 2011, p. 43). This 'history that never ends' includes more recent attacks on the form of communal land ownership that Aboriginal activists established with the passage of *Aboriginal Land Rights Act 1976* (NT) (ALRA), an Act that granted the lands occupied by 'reserves' to their Aboriginal owners and established a system whereby Aboriginal people could claim land. These attacks were both indirect and direct. Aboriginal communities have been chronically underfunded leaving them without healthcare, housing, or education infrastructure. The *Australian Medical Association Report Card 2007* on Aboriginal health, for instance, describes the government's failure to provide resources for Aboriginal people as 'criminal.' The report's authors argue that an additional AU$460m per year must be spent on Aboriginal healthcare, especially in the very programs abolished under the Intervention: community-controlled primary care (*Australian Medical Association Report Card Series 2007: Aboriginal and Torres Strait Islander Health* 2007, p. 6). As Irene Watson (2009) documents, the Howard Government also directly attacked the ALRA in 2006. Underpinned by their representation of Aboriginal forms of land tenure as 'threats' to 'economic security,' they amended this piece of legislation to open Aboriginal land up to commercial interests such as miners. They also laid the legislative framework for 99 year leasing agreements that aimed, in Brough's words, to 'provide for individual property rights' and home ownership in Aboriginal communities (2006, p. 5).

In the context of these repeated attempts to abolish communal land tenure, the racialised figure of the house must be conceptualised as constitutive of what da Silva (2009) identifies as the settler-colonial state's self-preserving force. Raciality courses through representations of the house, and coextensive notions of 'private property,' 'economic security,' and 'business management,' to (re)produce the legal fiction of *terra nullius*. On one level, race works within the politico-juridical rhetoric surrounding housing to (re)present targeted Aboriginal communities as international battlefields that exist in a premodern "state of nature." On another level, these figures also (re)produce post-9/11 biopolitical regimes of securitisation. Representations of Aboriginal communities as "wastelands" void of 'individual land title' resonate profoundly with the

War on Terror and the transnational imperial formations that predicate "global security" on the amelioration of "failed states." These representations are implicated within the state's compulsory acquisition of targeted Aboriginal land, use of military and police force, and plans to build "normal suburbs" and housing. Further, these practices are implicated in the 'logic of obliteration' that scripts targeted Aboriginal people's communal land tenure as a 'threat' that must be eliminated in the name of protecting white sovereignty (da Silva 2007).

'By Force of This Subsection': Settler-Colonial Self-Preservation

My point of entry into this colonial project lies in part four of the *Northern Territory National Emergency Response Act 2007* (Cth) (NTNER). Entitled 'Acquisition of rights, titles and interests in land,' two sections are particularly important. They state:

30A Object of Part

The object of this Part is to enable special measures to be taken to:

(a) improve the delivery of services in Indigenous communities …
(b) promote economic and social development in those communities

31 Grant of lease for 5 years

(1) A lease of the following land is, by force of this subsection, granted to the Commonwealth by the relevant owner of the land. (NTNER 2007, pp. 30–31)

Followed by over ninety pages of geographic data, which read like a catalogue of land to be annexed, these sections provide the Commonwealth with five-year leases over targeted Aboriginal land and thus the mandatory suspension of communal Aboriginal land title. They are closely tied to the Intervention's housing policy. Under these leases, the Commonwealth is not only entitled to 'exclusive possession' and 'quiet enjoyment' of the land, coupled with additional pieces of legislation, these leases also ensure that the state is entitled unrestricted access to the land—it can acquire all assets and maintain ownership of any houses that are built or repaired under the policy (NTNER 2007, p. 34). As the Central Land Council, a statutory authority comprised of 90 elected Aboriginal representatives who govern the southern half of the Northern Territory, wrote: 'No leases were negotiated ... these leases gave the Australian Government rights to exclusive possession, to repair, demolish or replace any existing building and infrastructure, and to unilaterally terminate the lease at any time' (Central Land Council 2013, pp. 14–5).

This legislative manoeuvre begins to expose how race operates through the Intervention and its legal infrastructure in the name of the settler-colonial state's self-preservation. Subsection 31 (1) is telling in this respect. In particular, the clause 'by force of this subsection' opens up a series of questions. Although it attests to its own force and legality, does this subsection have the power to (re)take possession of Aboriginal land? And is this act legitimate? How does it sustain its claim to legitimacy? According to Irene Watson, these questions not only haunt the NTNER legislation, they also haunt the state's everyday operations. She argues:

> A question the Australian state is yet to resolve is its own illegitimate foundation and transformation into an edifice deemed lawful. Within this unanswered questionable structure ... the survivors of this founding violence ask the state: by what lawful process do you come to occupy our lands? (2009, p. 46)

Subsection 30A (b) provides some answers to these questions. As it describes that the 'object of the part' is to 'promote economic and social development in those [Aboriginal] communities' (NTNER 2007, p. 30), the Act explicates Watson's thesis that the continuation of the colonial project is predicated on reproducing foundational moments of originary violence (2009). The use of the word 'development' is critical here. It illuminates historical continuities of settler colonialism and is racially charged. The usurpation of Indigenous sovereignty over country is always already legitimated by the (il)legal fiction of *terra nullius* that represented Indigenous country as void of 'development': as "unsettled" and without "civilised" human subjects equipped with systems of law, exchange, and property. With its proposal that Aboriginal land must be acquired in order to ensure 'development' then, the state (re)calibrates the racist representations that belie the state's legal foundations in accordance with contemporaneous expediencies; it (re)produces contemporary Aboriginal communities as void of economic and social structures. Further, these communities are represented as lacking the market networks that stand as signifiers of modernity and private property. Targeted Aboriginal land is represented as a "wasteland" awaiting colonisation and, as I will proceed to argue, is portrayed as a site that must be 'developed' by abolishing communal Aboriginal land title and establishing individual property rights and home ownership.

Howard also represents targeted Aboriginal communities as wastelands. Speaking on morning television the day after the Intervention was announced, he said: 'What we have got to do is confront the fact that [in targeted Aboriginal communities] the basic elements of a civilised society don't exist' (2007b). Here, Howard reproduces the violent fictions instantiated with white invasion. His assertion that the basic elements of civilisation do not 'exist' works to represent targeted Aboriginal people as primordial, lawless, and existing within a state of nature. White Australians, by implied contrast, are represented as self-determined subjects. Through his use of pronouns, Howard appeals to white Australians such as myself. He calls on "us" to 'confront' the need to take 'special measures' in targeted Aboriginal communities. With this move, he also works to construct "us" as benevolent and rational. Howard's use of the word 'confront' infers that, although 'we' may prefer to be compassionate and to forgo judgement, 'we' 'must confront the fact[s]' and take reasoned action. These representations culminate to (re)produce *terra nullius* in the contemporary context of the Intervention. They work in concert with sections 30A and 31 of the NTNER Act to (re)assert white possession of unceded Indigenous country.

The figures of the house and private property are significant in this context. Brough uses the construction and 'improvement' of housing to justify the compulsory (re)acquisition of targeted Aboriginal land. In order to elucidate this operation, I want to draw on da Silva's (2001, 2007, 2009) groundbreaking reconceptualisation of race. According to da Silva, race cannot be conceptualised as a mechanism that invokes "difference" in order to produce, and later justify, the exclusion of target subjects from human rights and attendant legal frameworks. It must be understood as a politico-symbolic arsenal that unravels the distinctions between race, law, and sovereignty. The product of knowledge/power, she proposes that raciality always already operates through the state's architecture in order to reproduce whiteness as universal. As she argues, 'whiteness [has] been produced to signify the principles of universal equality and freedom informing our conceptions of the Just, the Legal and the Good' (da Silva 2001, p. 423). Raciality, then, is productive. It produces whiteness—and white bodies—as always already lawful, ethical, self-determined, and situated within the powerful domain of universality. Raciality also necessitates specific modes of academic inquiry. As da Silva continues: 'I believe only by examining how the racial has produced the domain, *universality*, will it be possible to work towards the enlargement of the horizons of Justice' (2001, p. 423; original emphasis).

Coupled with her approach to studies of race injustice, da Silva's (2001, 2007, 2009) understanding of race explicates Brough's discussion of the NTNER Act. It demonstrates how raciality works as an *a priori* within colonial law and representations of housing in order to legitimate the settler-colonial state's authority. In the midst of hurried debate that facilitated the passage of this legislation through the House of Representatives in just one day, Brough explained: 'Currently, there are too few jobs in these communities and land tenure arrangements work against developing a real economy' (2007c, p. 11). He continued: 'when land tenure is settled, the Howard Government will begin the process of improving housing and infrastructure dramatically' (2007c, p. 15). Although he omits racial signifiers in these statements, Brough's words are neither neutral nor anti-racist. On one level, they are coercive. He outlines that the provision of infrastructure, such as houses and medical facilities, is contingent on the acquisition of targeted Aboriginal land. On another—interdependent—level, his references to 'land tenure,' the 'real economy,' 'housing,' and 'settle[ment]' are also telling. They extend Howard's depiction of targeted Aboriginal communities as *terra nullius*. In a move that exemplifies da Silva's (2001) thesis, Brough represents colonial law and the neoliberal market as 'transparent,' universal, and ethical. For instance, his presupposition that Aboriginal land and town camps, governed by communal title, are void of 'settled' 'land tenure' is significant. It (re)produces the analytics of raciality that always already represent whiteness as universal, civilised, and property bearing. With this move, then, Brough universalises colonial forms of land tenure and simultaneously renders Aboriginal country "un-settled" and "un-developed" "waste." His claim that communal land title 'work[s] against developing a real economy' reinforces this racialised representation. In line with section 30A (b) of the Intervention that suggests the compulsory leases will promote 'economic development,' Brough's use of the adjective 'real' reproduces what da Silva and Chakravartty (2012) identify as the 'racial logic of capitalism.' It not only posits that capitalism is universal, it also suggests that white Australia's land tenure is "valid" and that individual home ownership drives the neoliberal market.

'Stabilise, Normalise and Exit': The Intervention as National Security

Brough's call for targeted Aboriginal communities to enter the 'real economy' also brings into focus the post-9/11 regimes of securitisation that are reproduced by the Intervention. Building on the characterisation of targeted Aboriginal communities as 'nothing less than war zone[s]' and 'failed societies,' his language frames this policy as an incursion into an external battlefield conducted in the name of securing white sovereignty (Brough 2007a, p. 10, c). On the day of the Intervention's announcement, for instance, he told the House of Representatives that: 'there are three phases to what we are doing: (1) stabilisation, (2) normalisation and (3) exit' (Brough 2007b, p. 77). Here, Brough (re)produces the analytics of raciality that always already position Aboriginal bodies, minds, and territories outside the domain of universality. He not only positions targeted Aboriginal communities as unstable and 'affectable' spaces within which the state must impose the colonial rule of law and install the neoliberal market place; he also asserts that the white state's forces must enter and reappropriate such racialised zones in order to quell the 'threat' they pose to authority (da Silva 2009). As da Silva writes:

[R]aciality produces both the subject of ethical life, who the halls of law and forces of state protect, and the subjects of *necessitas*, the racial subaltern subjects whose bodies and territories, the global present, have become places where the state deploys its forces of self preservation. (2009, p. 224)

In the context of the Intervention, Howard and Brough's use of war-like rhetoric explicates this operation. For instance, Howard's announcement that his Government will take the 'hardline approach … we are moving in, we are going to take control' represents his policy as a militaristic manoeuvre (Howard 2007b). As his repetitive use of the phrase 'we are' continues to (re)produce white Australians as self-determined and lawful, it suggests that 'we are' staging an incursion into—or a "take over" of—a foreign territory. Brough extends this representation, claiming: 'this is a great national endeavour. It is the right thing to do, and now is the right time to do it' (Brough 2007c, p. 17). Building on Howard's rhetoric, he utilises a nationalistic narrative to posit that the settler-colonial state is forcibly—and nobly—entering and fighting within foreign and danger-ous territory.

This representation highlights the critical continuities between the Intervention and the War on Terror. Howard and Brough's use of material force evidences the way in which raciality represents targeted Aboriginal communities as external theatres of war populated by potential threats to white security. Over 600 additional police, military, and investiga-tive personnel from the Australian Crime Commission were mobil-ised under Operation Themis, Operation Outreach, and the National Intelligence Taskforce into Indigenous Violence and Child Abuse, respec-tively (Snowdon 2008). The name of the military's mission, 'Operation Outreach,' (re)enforces Howard and Brough's war-like apparatus. It represents white Australians as powerful soldiers working outside white Australia's borders. As Suvendrini Perera writes, 'in the context of the Global War on Terror, inside and outside become intersecting domains for the staging and reaffirmation of Australia as a white nation and a launching ground for renewed missions of racial salvation' (2009, p. 119). Perera's understanding of Australia's role in the War on Terror explicates the way in which Howard and Brough redraw the state's borders in order to represent targeted Aboriginal communities as beyond the reaches of universality. Relegated outside the settler-colonial state's boundaries, they are reproduced as war zones that Australian personnel must secure. The testimony of Donald Kunoth, vice president of the Kunoth Town Camp, exposes the extra-discursive ramifications of this representation. Describing a police raid after the commencement of the Intervention, he said: 'There was a big mob of police here. They came running in like they were looking for terrorists. We've never had that before' (cited in Gibson 2012, p. 88).

'We Have Lived in This Country as a Foreigner!': Aboriginal Economic Frontiers

Alongside the deployment of the state's material forces into targeted Aboriginal communities, the construction of housing and infrastructure must also be understood as constitutive of the settler-colonial state's self-preserving force. Under the Intervention, these projects are represented as moves designed to extend the Australian state's economy and control into insecure "foreign" territory. Following the compulsory acquisition of targeted Aboriginal communities, for instance, the Howard Government immediately dissolved the community-owned organisations charged with building Aboriginal housing and acquired their assets (NTNER 2007). It established the Strategic Indigenous Housing and Infrastructure Program (SIHIP) and, in 2008, announced that the construction of Aboriginal people's homes would be outsourced to external contractors. As the Northern Territory's Minister for Housing, Kon Vatskalis stated in his media release: 'About 150 construction industry representatives will gather in Darwin today to find out how they can share in the $647 million landmark housing program' (Vatskalis 2008, p. 1). In keeping with the racial logic of capitalism, Vatskalis's reference to companies 'sharing' AU$647m is significant. It represents targeted Aboriginal communities as an economic frontier. They are posited as 'new territories of invest-ment' within which business and the state can expand and extract profit (Chakravartty and da Silva 2012). This representation is compounded by the Commonwealth's decision to appoint Parsons Brinckerhoff as project managers (Vatskalis 2008, p. 1). A multinational engineering company,

Parsons Brinckerhoff operate across the world designing and constructing mining and infrastructure programs. Notably, they also work in collaboration with the US Military and Department of State in the reconstruction of Iraq (Parsons Brinckerhoff 2015). Reproducing the positioning of targeted Aboriginal communities as "war zones," then, Aboriginal lands are also represented as dangerous sites that must "open up" to foreign investment in the name of Western security. This discourse is further reflected within the names of two of the 'Strategic Alliances' that operate in the Northern Territory. Formed in order to win the contracts to build Aboriginal housing, these "alliances" are comprised of many construction and infrastructure companies. They too position themselves on the economic frontier. Called the 'Earth Connect' and 'New Future' Alliances, they are represented as connecting with undeveloped "earth" in order to build the modern infrastructure that will secure the state's "new future" (Auditor-General for the Northern Territory 2010, p. 11).

The project to build housing and a "new future" in targeted Aboriginal communities points to another regime of biopolitical securitisation that is important here. It can be linked to the way the Commonwealth effectively seized the assets, funds, and housing owned by each targeted Aboriginal community and, in turn, appointed GBMs to manage them (NTNER 2007). These measures invoke both historical and geopolitical points of connection. In the first instance, the sections of the NTNER Act devoted to empowering GBMs echo key sections of the *Northern Territory Aboriginals Protection Act* of 1910 that gave the Chief Protector of the Territory extensive powers to manage Aboriginal people's everyday lives, including their finances, wages, and estates. In particular, it echoes one section that gave the Chief Protector the power to 'take possession of, retain, sell or dispose of and give a valid title to any such property [within the Aboriginal reserve], whether real or personal' (*Northern Territory Aboriginals Act* 1910, p. 12). Under Howard and Brough, this list of material possessions is extended to include incorporated Aboriginal organisations, community-owned services, and Aboriginal housing.

The Howard Government also extends the biopolitical role of white 'Chief Protectors.' The 'Government Business Managers Statement of Roles and Responsibilities,' for example, states that managers are 'the single face of the Australian Government at the local community level—akin to an ambassador' (cited in the Australian National Audit Office 2010). The use of the word 'ambassador' in this statement represents managers as representatives of the Australian Government working in foreign countries. Galiwin'ku lawman Rev Dr Djiniyini Gondarra OAM (Order of Australia) identifies this colonial arsenal. In a book published to share the voice of Indigenous Elders across the country, he said: 'Don't let the other people, the First People of this country, be rejected! Being seen as the second-class citizen! Being seen as the outcast! We have lived in this country as a foreigner!' (cited in Harris 2011, p. 45). Here, Gondarra identifies the way in which targeted Aboriginal people and their systems and laws are represented as foreign and regulated from within the state's legislative borders where their cultures and forms of land tenure can be, in Foucauldian terms, 'let to die.' In this context, then, GBMs are represented as those who must obliterate communal land title and organisations. Moreover, they are tasked with "normalising" Aboriginal communities in the name of "protecting" and "securing" the interests of the white Australian populace.

In an opinion piece in the *Australian*, historian John Hirst applauds the imposition of 'privatising' land tenure and business managers. Following his argument that 'Australia is a long-established capitalist society,' he continues: 'The [Aboriginal] people have shown they are incapable of governing themselves. There is no point in consulting them about the creation of authority; authority has to be created for them' (2007, p. 12). Here, Hirst effaces scholarship that documents the long-standing economic and trade relationships between Aboriginal and Torres Strait Islander communities and the wider Asia-Pacific in order to sustain the "White Protector" mythology (Balint 1999; Perera 2009). He exemplifies the ways in which white Australians depict targeted Aboriginal people, in line with Gondarra's analysis, as animal-like "others" who are lesser than white Australians and "lack" the self-determination necessary to exist within contemporary economies (Chakravartty and da Silva 2012). Coupled with the depiction of GBMs as 'ambassadors,' Hirst reproduces biopolitical regimes of power that not only represent targeted Aboriginal minds, bodies, and territories as "criminal" but also "deviant" "threats." Like Chief Protector Cook before them, Howard, Brough, and Hirst script target subjects as bodies that can be lawfully "normalised" and obliterated under the auspices of preserving the settler-colonial state.

'We're Saying No': Indigenous Resistance to the Settler-Colonial Project

As I write this conclusion, the settler-colonial state's self-preserving force ramifies into the present. Howard and Brough's policies have been (re) instantiated by the Rudd, Gillard, and Abbott Governments that followed. The state's economic and juridical infrastructure consistently attempts to (re)assert, legitimate, and secure white sovereignty. For instance, the 'secure tenure' regime holds many targeted Aboriginal communities to ransom (Central Land Council 2013). Under this regime, the Government will not provide infrastructure funding for housing unless they sign over their land to the Government on a 40–99 year lease and work towards replacing communal land title with individual home ownership and private property.

Indigenous activists have demonstrated how this more recent iteration of the ongoing settler-colonial project reproduces foundational colonial violence. Kunoth-Monks told reporters, 'Right now we are again traumatised because [the land is] the last stable thing we feel under our feet: our earth, our ground, *our home* of thousands of years' (cited in Special Broadcasting Service 2013; emphasis added). Here, Kunoth-Monks not only attests to her unceded sovereignty over her country, her earth, and her home, she also demonstrates that the racial logic of capitalism has interlocked with colonial law in the name of "home ownership" to once again script the obliteration of Aboriginal sovereignty as "necessary." The Alyawarr people who built and occupy Protest House have also rejected the settler-colonial state's colonising arsenal. As Protest House asserts their sovereignty over country, it challenges the ways in which the Intervention continues to (re)instantiate the legal fiction of *terra nullius* that belies the white settler-colonial state's legitimacy. As spokesman Richard Downs states: 'We're a remote community. Our traditions and customs are still strong, our law's still strong, Aboriginal way. We're not going to let it go away' (cited in Graham 2009). He continues: 'the government is playing a waiting game. It thinks we'll get sick of it and go back to the community. But we're saying no. We're never ever going to go back to that community to live under your controls and measures' (Downs 2009a).

Note

1. Established in the late 1970s, CDEP was a scheme that provided block grants to community-controlled organisations so that they could employ local Aboriginal people on flexible contracts to complete work within the local community, such as driving the school bus and developing communal infrastructure (see Altman 2007).

References

Adlam, N., & Gartress, A. (2007, June 22). Martial law. *Nothern Territory News*, p. 1, 7.

Altman, J. (2007). *Neo-paternalism and the destruction of CDEP*. Canberra: Centre for Aboriginal Economic Policy Research.

Auditor-General for the Northern Territory. (2010). *Strategic indigenous housing and infrastructure program June 2010 report to the Legislative Assembly*. Darwin: Northern Territory Government.

Australian Broadcasting Corporation. (2007, June 26). Lateline transcript: Govt orchestrating land grab. http://www.abc.net.au/lateline/content/2007/s1962845.htm. Accessed 4 June 2014.

Australian Human Rights and Equal Opportunities Commission. (1997). *Bringing them home: Report of the national inquiry into the separation of Aboriginal and Torres Strait Islander children from their families*. Canberra: Commonwealth of Australia.

Australian Medical Association Report Card Series 2007: Aboriginal and Torres Strait Islander Health. (2007). Australian Medical Association.

Australian National Audit Office. (2010). *Government business managers in aboriginal communities under the northern territory response*. http://www.anao.gov.au/Publications/Audit-Reports/2010-2011/Government-Business-Managers-in-Aboriginal-Communities-under-the-Northern-Territory-Emergency-Response/Audit-brochure. Accessed 20 Dec 2014.

Balint, R. (1999). The last frontier: Australia's maritime territories and the policing of Indonesian fisherman. *Journal of Australian Studies, 23*(63), 30–39.

Bowlby, S., Gregory, S., & McKie, L. (1997). "Doing home:" Patriarchy, caring and space. *Women's Studies International Forum, 20*(3), 345–350.

Brough, M. (2006). *House of Representatives Hansard: 31 May 2006.*

Brough, M. (2007a, October 2). *Alfred Deakin Lecture at Melbourne University.* http://pandora.nla.gov.au/pan/36764/20071026-1200/www.facs.gov.au/internet/minister3.nsf/content/alfred_deakin_02oct07.htm. Accessed 9 Mar 2011.

Brough, M. (2007b). *House of Representatives Hansard: 21 June 2007.*

Brough, M. (2007c). *House of Representatives Hansard: 7 August 2007.*

Central Land Council. (2013). *Land reform in the Northern Territory: Evidence not ideology.* http://www.clc.org.au/publications/content/land-reform-in-the-northern-territory-paper. Accessed 10 Sept 2014.

Chakravartty, P., & da Silva, D. F. (2012). Accumulation, dispossession, and debt: The racial logic of global capitalism—An introduction. *American Quarterly, 64*(3), 361–385.

da Silva, D. F. (2001). Towards a critique of the socio-logos of justice: The analytics of raciality and the production of universality. *Social Identities, 7*(3), 421–454.

da Silva, D. F. (2007). *Toward a global idea of race.* Minneapolis: University of Minnesota Press.

da Silva, D. F. (2009). No-bodies: Law, raciality and violence. *Griffith Law Review, 18*(2), 212–236.

Downs, R. (2009a, October 7). *Media releases.* https://interventionwalkoff.wordpress.com/media-releases/. Accessed 1 Nov 2014.

Downs, R. (2009b). *Stop the NT Intervention—Support the Ampilatwatja Walk Off.* https://interventionwalkoff.files.wordpress.com/2009/08/090823_ampilatwatja_supportletter1.pdf. Accessed 4 Nov 2014.

Foucault, M. (2004). *"Society must be defended:" Lectures at the college De France 1975–76.* London: Penguin.

Fox, L. (2008). "Re-possessing home:" A Re-analysis of gender, homeownership, and debtor default for feminst legal theory. *William and Mary Journal of Women and the Law, 14*(2), 243–294.

Gibson, P. (2012). Return to the ration days: The northern territory intervention—grass roots experience and resistance. *Ngiya: Talk the Law, 5*(3), 58–107.

Goodall, H. (1995). "Assimilation begins in the home:" The state and Aboriginal women's work as mothers in New South Wales, 1900s to 1960s. *Labour History, 69*, 75–101.

Graham, C. (2009, August 5). Up to their ankles in sewage, a remote community's patience runs out. *Crikey.* http://www.crikey.com.au/2009/08/05/

up-to-their-ankles-in-sewage-a-remote-communitys-patience-runs-out/. Accessed 1 Nov 2014.

Harris, M. (2011). *Walk with us*. Melbourne: Concerned Australians.

Hirst, J. (2007, June 26). The myth of a new paternalism. *The Australian*, p. 12.

Howard, J. (2007a, June 25). *Address to the Sydney Institute, Four Seasons Hotel, Sydney*. http://pandora.nla.gov.au/pan/10052/20080118-1528/pm.gov.au/media/Speech/2007/Speech24394.html. Accessed 3 Mar 2011.

Howard, J. (2007b, June 22). *Interview with David Koch and Melissa Doyle Sunrise, Seven Network*. http://pmtranscripts.dpmc.gov.au/release/transcript-15653. Accessed 1 Sept 2011.

Howard, J. (2007c, June 21). *Joint press conference with the Hon Mal Brough, Minister for Families, Community Services and Indigenous Affairs*. http://pandora.nla.gov.au/pan/10052/20080118-1528/pm.gov.au/media/Interview/2007/Interview24380.html. Accessed 1 Feb 2011.

Karvelas, P. (2007, June 22). Crusade to save Aboriginal kids. *The Australian*, p. 1.

Korff, J. (2015, April 7). Aboriginal history timeline (1900–1969). *Creative Spirits*. http://www.creativespirits.info/aboriginalculture/history/aboriginal-history-timeline-1900-1969. Accessed 20 Apr 2015.

Kunoth-Monks, R. (2012). NT Intervention Forum—Rosalie Kunoth-Monks. *Arena*. http://arena.org.au/nt-intervention-forum-rosalie-kunoth-monks/. Accessed 10 Dec 2013.

Markus, A. (1990). *Governing savages: The Commonwealth and Aboriginies 1911–1939*. Sydney: Allen & Unwin.

Parsons Brinckerhoff. (2015). Iraq power sector reconstruction. https://pbworld.com/capabilities_projects/iraq_power_sector_reconstruction.aspx. Accessed 9 Mar 2015.

Perera, S. (2009). *Australia and the insular imagination: Beaches, borders, boats and bodies*. New York: Palgrave Macmillan.

Rothwell, N. (2007, June 22). Nothing less than a new social order. *The Australian*, pp. 1–4.

Snowdon, W. (2008, October 31). *Successful conclusion to operation outreach*. http://www.defence.gov.au/minister/74tpl.cfm?CurrentId=8397. Accessed 1 Apr 2015.

Special Broadcasting Service. (2013). Former Prime Minister Malcolm Fraser has criticised a federal government push to introduce 99-year leaseholds over some Northern Territory communities, 27 November. http://www.sbs.com.au/news/article/2013/11/27/malcolm-fraser-criticises-governments-leasehold-plan. Accessed 12 Dec 2014.

Vatskalis, K. (2008, April 30). Construction Industry to Learn More About Strategic Alliances. *Northern Territory Government Media Release.*
Watson, I. (2009). What is saved or rescued and at what cost in the northern territory intervention. *Cultural Studies Review, 15*(2), 45–60.

Legislation

Aboriginal Land Rights Act 1976 (NT).
Northern Territory Aboriginals Ordinance 1911 (Cth).
Northern Territory Aboriginals Act 1910 (SA).
Northern Territory National Emergency Response Act 2007 (Cth).

Erratum: Domesticating Drone Technologies: Commercialisation, Banalisation, and Reconfiguring 'Ways of Seeing'

Caitlin Overington and Thao Phan

The original version of the book contained an error which has been corrected.

The correction is given below:

Chapter 8

The author names were listed incorrect in the original book. The correct order is given below:

C. Overington
School of Culture and Communication,
University of Melbourne, Melbourne, VIC, Australia

T. Phan
School of Social and Political Sciences,
University of Melbourne, Melbourne, VIC, Australia

The updated original online version for this chapter can be found at DOI…
http://dx.doi.org/10.1057/978-1-137-55408-6_8

© The Author(s) 2016   E1
H. Randell-Moon, R. Tippet (eds.), *Security, Race, Biopower*,
DOI 10.1057/978-1-137-55408-6_11

Conclusion

On 15 September 2012, a Welcome to Aboriginal Land Passport Ceremony took place in the traditional lands of the Gadigal people in Sydney, Australia (Aboriginalpassportceremony.org 2012). Exercising their sovereign rights to welcome and care for new migrants, Robbie Thorpe from Treaty Republic and an elder from the Indigenous Social Justice Association (ISJA) issued the passports to over two hundred people including refugees and asylum seekers in absentia, incarcerated as part of Australia's mandatory detention policy.[1] The elder from ISJA explained the purpose of the 2012 ceremony to those issued with the passports, 'Whilst they acknowledge our rights to all the Aboriginal Nations of Australia we reciprocate by welcoming them into our Nations' (cited in Pugliese 2015, p. 88). Passport ceremonies, such as this one, reveal how the sovereign responsibilities of some Indigenous communities continue to operate alongside and in defiance of the settler state's claim to authority over its borders and the peoples that move within it. The ceremony forms part of a history of sovereign practices where Indigenous peoples have issued their own protocols and documentation for travelling and recognising belonging within and outside their country. Such practices expose the multiplicity and incommensurability of

© The Author(s) 2016
H. Randell-Moon, R. Tippet (eds.), *Security, Race, Biopower*,
DOI 10.1057/978-1-137-55408-6

sovereignties and laws ignored by dominant geopolitical formations of state-sanctioned border policing.

As we write this conclusion, the numbers of "state-less" persons moving across borders is occurring at historically unprecedented numbers. Nation-states are scrambling to regulate this migration through a bio-politics of quotas that would permit "acceptable" levels of new citizens. Such calculations are exemplary of the apparatuses of security, race, and biopower examined in this book. The labelling of migrations from Syria, in particular, as a refugee "crisis" obfuscates a crisis of state security that condones warfare and foreign military intervention, ostensibly to secure for the region the liberal freedoms enjoyed by citizens in the Global North. Yet this same liberality is economised as finite, as needing to be apportioned carefully to the bodies fleeing the violence of regimes whose apparent failures necessitated intervention in the first place. Indigenous responses to refugee migrations emerge from a complex history and entanglement with laws and states that are the product of invasive migrations. These invasive migrations form part of the imperial and colonial histories that underprop contemporary state surveillance, which find expression in the violence applied to bodies that threaten the maintenance of territorial integrity. Indigenous passport ceremonies are made possible by the continuation and development of lore that does not require the historical contingency of the settler-state and its governmental bodies for viability.

This collection has focused on the contemporary geocorpographies of bodies, space, and technology: where technologies have revealed new possibilities for identifying and locating risk in particular bodies in particular spaces. We have argued that contemporary racisms and processes of racialisation instrumentalise an older aleatory logic that both requires and justifies the production of disposable bodies in the development and testing of technologies of law, war, and medicine. Following the work of Foucault, we have suggested that this testing and application of geopolitical risk finds its surface of intervention on bodies because they are the corporeal signifier of an identity. It is because bodies express identity that technologies can sort them into appropriate spaces and places of "value".

But life exists beyond the body, and resistance to technological and spatialised instruments of power occurs—in surreptitious, unexpected, and unflinching ways. Resistance is exercised through commitment to country and the wellbeing of its people, as with the Ampilatwatja walk-off and refusal to recognise the Australian government's Northern Territory Intervention; the "irresponsible" (because erotic) consumption of life-saving drugs such as pre-exposure prophylaxis (PrEP); and the birth of a prince interrupting Commonwealth narratives of gender equality. Meanwhile, the use of health apps and smart devices are designed to focus attention on the body's health and capillary rhythms, encouraging a continuous disciplining of the self. This attention to the self may produce what Foucault describes as 'a counter-attack in that same body' by revealing capacities and capabilities hitherto unforeseen by strategies of power.

> Mastery and awareness of one's own body can be acquired only through the effect of an investment of power in the body … But once power produces this effect, there inevitably emerge the responding claims and affirmations, those of one's own body against power, of health against the economic system, of pleasure against the moral norms of sexuality, marriage, decency. Suddenly, what had made power strong becomes used to attack it. (1980, p. 56)

Following a Foucauldian approach to embodied power reveals how the capacity for resistance is built into structures of governance which presume, as their bases for efficacy, the freedom of subjects to submit; 'without the possibility of recalcitrance, power would be equivalent to a physical determination' (Foucault 1994, p. 342). The assumption that elderly residents of nursing homes will leave, for instance, requires surveillance and regulation; the space of the airport produces unexpected affective and somatic encounters and so must be tightly regulated; and the use of drones for "leisure" purposes encourages new 'ways of seeing'. It is precisely because of 'the intransigence of freedom' (p. 342) that Foucault views the conditions of knowledge that constitute bodies as recognisable subjects as: 'war and battle. The history which bears and determines us has the form of a war rather than of a language: relations of power, not relations of meaning' (Foucault 1980, p. 114). For Foucault then, conflict over competing knowledge claims and the differing degrees to which bodies are subject and make themselves subject to apparatuses of power form 'a permanent provocation' (Foucault 1994, p. 342).

Such provocations centre on the historical refusal of individuals and groups to be abstracted by 'economic and ideological state violence, which ignore who we are individually' because of the 'scientific or administrative inquisition that determines who one is' (p. 331) and one's utility to the state and social body. Resistance, then, is constitutive and an effect of the various ways power subjectifies individuals and groups.

For Foucault, the individual is always subject to and a subject of power. Are there circumstances in which subjectivity—to be a subject—is not possible? Whilst recognising the possibilities for agency and resistance within power structures we must not forget Franz Fanon's contention that those who bear 'the burden of ... corporeal malediction' (2008, p. 84) are made 'objects' (p. 85) rather than subjects of their histories. Articulating how 'whiteness burns me', Fanon describes his quotidian encounters with white people as 'the field of battle having been marked out' (p. 86). In this theatre of battle, some bodies and their consciousness are set for destruction so that others may exist in safety and comfort.

In Section One: Geocorpographies, Joseph Pugliese's chapter opened the book with an account of how drone technologies create the bodies of people in Yemen and Pakistan as objects of war, reducible to the flesh and minerals of their surrounding environs. Joshua Pocius considered the different types of consumption for PrEP based on differently located and risk-assessed bodies in the Global North and Global South. The use of secular law to preserve the white, Anglican, and heteronormative bodies of the British monarchy to keep intact the integrity of the Commonwealth was the focus of Holly Randell-Moon's chapter. Finally, Sunshine Kamaloni examined the airport as a geocorpography that is generative of racialised affect and surveillance to keep black bodies in place.

Section Two: Technologies outlined how smart technologies and the Internet are differently entwined in the life-cycle of labouring bodies. Ryan Tippet discerned in the Internet.org charity led by Facebook an expansion of the 'digital enclosure' for capitalising on the immaterial labour of mobile phone users. Brett Nicholls took aim closer to home, at the wearable health motivation technologies which epitomise a control society strategy of population maintenance. And for Sy Taffel, the production and distribution of smart digital technologies represented an under-examined cycle of environmental and corporeal damage, concealed under the rhetoric of "weightless" and "green" techno-utopianism.

The book concluded with an explicit investigation of the political strategies used by governing authorities and market forces to foster certain lives as productive and healthy whilst neglecting others. In this final section, Biopolitics, Caitlin Overington and Thao Phan discussed the extension of theatres of war into urban spaces through the emergence of drones as hobby and surveillance as leisure activity. David-Jack Fletcher asked poignant questions about the role of nursing homes and their facilitation of the removal of the elderly as desirable for the social body. In the final chapter's account of the Northern Territory (NT) Intervention, Jillian Kramer critically examined the Australian state's attempt to punish Aboriginal residents of remote NT communities for failing to adhere to a normal, property-owning subjectivity alongside the rejection of the state's "help" by the Alyawarr people—who see the Intervention's violent intrusions into everyday life as another iteration of a colonising mindset.

In the context of global migrations, border-surveillance, and the information revolution, we see how a geopolitics anchored in the territorial preservation of national borders, mutates and fragments as a dominant security paradigm. This book has attempted to map the heterogeneous modes of locating, cultivating, targeting, curing, killing, moulding, and resisting technologies of space and race. Across these chapters, an important reminder emerges: it is essential that in the defence of the marginalised and oppressed, we remain critical of technologies and social formations that seek to categorise, compartmentalise, and hierarchise groups according to their essential characteristics.

Note

1. Under the *Migration Reform Act 1992* (Cth), persons who arrive by boat seeking asylum in Australia are subject to mandatory detention. Initially implemented as an interim measure to stem arrivals from Cambodia, the law has become a permanent part of the Australian state's asylum seeker deterrence policy, which also includes detention debts (that charge asylum seekers for their imprisonment) and the

Pacific Solution (where asylum seekers are housed in processing centres in the Pacific, away from the Australian mainland). For more information on the horrific conditions asylum seekers are subject to in these detention centres, see *Researchers Against Pacific Black Sites*: http://researchersagainstpacificblacksites.org.

References

Aboriginalpassportceremony.org. (2012, 7 August). More than 200 migrants to receive Aboriginal passports. *Green Left Weekly.* https://www.greenleft.org.au/node/51810. Accessed 22 Mar 2016.

Fanon, F. (2008). *Black skin, white masks* (Trans. C. L. Markmann). London: Pluto Press.

Foucault, M. (1980). *Power/knowledge: Selected interviews and other writings (1972–1977)* (Trans. Gordon, C. Marshall, L. Mepham, J. & Soper, K. Ed. Gordon, C). New York: Pantheon Books.

Foucault, M. (1994). The subject and power. In J. D. Faubion (Ed.), *Michel Foucault: Power*. New York: The New Press.

Pugliese, J. (2015). Geopolitics of aboriginal Sovereignty: Colonial law as "a species of excess of its own authority", Aboriginal passport ceremonies and asylum seekers. *Law Text Culture, 19*, 84–115.

Index

Note: Page numbers followed by n denote notes.

© The Author(s) 2016
H. Randell-Moon, R. Tippet (eds.), *Security, Race, Biopower*,
DOI 10.1057/978-1-137-55408-6

Printed in the United States
by Baker & Taylor Publisher Services